HEADLINE NEWS, SCIENCE VIEWS

Edited by
DAVID JARMUL
National Research Council

National Academy of Sciences
National Academy of Engineering
Institute of Medicine
National Research Council

NATIONAL ACADEMY PRESS
Washington, D.C.   1991

NATIONAL ACADEMY PRESS • 2101 Constitution Avenue, N.W. • Washington, D.C. 20418

The National Academy of Sciences is a private, nonprofit, self-perpetuating society of distinguished scholars engaged in scientific and engineering research, dedicated to the furtherance of science and technology and to their use for the general welfare. Upon the authority of the charter granted to it by the Congress in 1863, the Academy has a mandate that requires it to advise the federal government on scientific and technical matters. Dr. Frank Press is president of the National Academy of Sciences.

The National Academy of Engineering was established in 1964, under the charter of the National Academy of Sciences, as a parallel organization of outstanding engineers. It is autonomous in its administration and in the selection of its members, sharing with the National Academy of Sciences the responsibility for advising the federal government. The National Academy of Engineering also sponsors engineering programs aimed at meeting national needs, encourages education and research, and recognizes the superior achievements of engineers. Dr. Robert M. White is president of the National Academy of Engineering.

The Institute of Medicine was established in 1970 by the National Academy of Sciences to secure the services of eminent members of appropriate professions in the examination of policy matters pertaining to the health of the public. The Institute acts under the responsibility given to the National Academy of Sciences by its congressional charter to be an adviser to the federal government and, upon its own initiative, to identify issues of medical care, research, and education. Dr. Samuel O. Thier is president of the Institute of Medicine

The National Research Council was organized by the National Academy of Sciences in 1916 to associate the broad community of science and technology with the Academy's purposes of furthering knowledge and advising the federal government. Functioning in accordance with general policies determined by the Academy, the Council has become the principal operating agency of both the National Academy of Sciences and the National Academy of Engineering in providing services to the government, the public, and the scientific and engineering communities. The Council is administered jointly by both Academies and the Institute of Medicine. Dr. Frank Press and Dr. Robert M. White are chairman and vice chairman, respectively, of the National Research Council.

Library of Congress Cataloging-in-Publication Data

Headline news, science views / David Jarmul, editor; National Academy
  of Sciences, National Academy of Engineering, Institute of Medicine,
  National Research Council.
        p.   cm.
    Includes index.
    ISBN 0-309-04480-4: $24.95. — ISBN 0-309-04384-0 (pbk.): $14.95
    1. Science news —United States. 2. Science —Social aspects—
United States. 3. Technology —Social aspects—United States.
I. Jarmul, David. II. National Research Council (U.S.)
    Q225.H43   1991
303.48'3—dc20   91-7480
                                                                    CIP

Copyright © 1991 by the National Academy of Sciences

No part of this book may be reproduced by any mechanical, photographic, or electronic process, or in the form of a phonographic recording, nor may it be stored in a retrieval system, transmitted, or otherwise copied for public or private use, without written permission from the publisher, except for the purposes of official use by the U.S. government.

Printed in the United States of America

Contents

Frank Press	Foreword	ix
	Editor's Note	xiii

1
SCIENCE AND NON-SCIENTISTS

Bill Cosby	Getting the Facts Straight About Science	3
Steven L. Goldman	Who Killed Yankee Ingenuity?	6
Ben Patrusky	On an Antidote for Science Phobia	9
Leon M. Lederman	Physics for Poets, Science for Society	11
John Ahearne	Making Sense of a Risk-Filled World	14
Thomas H. Kean	Making the Link Between Science and Politics	16

2
TECHNOLOGY IN EVERYDAY LIFE

Charley V. Wootan	Making Our School Buses Safer 21
Samuel H. Fuller and Damian M. Saccocio	A Computer Future Without a Heart 24
Robert D. Ervin and Kan Chen	Toward Motoring Smart 27
Joseph M. Sussman	Easing the Crunch at Our Airports 30
John C. McDonald	Protecting Our Phones from Terrorism 33
Robert B. Kurtz	The New Arsenal of Democracy 35
Ezra Ehrenkrantz	Building Houses People Can Afford 39
Sara J. Czaja	Designing for an Aging America 41
Richard E. Hallgren	Preparing for the Next Big Natural Disaster 44
Robert F. Jortberg	Our $1.5 Trillion Investment 46
Robert G. Dean	Tough Choices About Rising Sea Level 48

3
A SUSTAINABLE FUTURE

Robert M. White	Uncertainty and the Greenhouse Effect 55
John J. Magnuson	Saving Sea Turtles 57
Hugh Downs	Who Owns Antarctica? 60
Nina Fedoroff	Genetically Engineered Organisms: Monsters or Miracles? 62
Frank L. Parker	Rethinking Radioactive Waste Disposal 65

CONTENTS

John Pesek	Toward a Sustainable Agriculture 67
Michael R. Taylor and Charles M. Benbrook	The Paradox of Pesticides 69
Jan van Schilfgaarde	Agriculture and Water Quality 72
James D. Nations	Exploring the Mysteries of 'Deep Ecology' 74

4

THE NATION'S HEALTH

Arno G. Motulsky	Food and Health 81
John C. Greene	The Spitting Image: Baseball Players and Chewing Tobacco 83
Robert D. Sparks	Clearing Our Vision About Alcohol Abuse 86
Howard W. Jones, Jr.	Needless Infertility 88
Roger J. Bulger	Who Is Going to Deliver Baby? 91
Susan S. Gallagher	Accidents Are Not Always Accidental 94
Heather Miller and Marshall Becker	Changing Behavior to Limit the Spread of AIDS 96
Robin Weiss and Theodore Cooper	The Dilemma of AIDS Drug Experiments 99
Samuel O. Thier	Identifying What Works in Medicine 101

5

MAKING SENSE OF SOCIAL PROBLEMS

Gerald David Jaynes and Robin M. Williams, Jr.	Confronting the Facts in American Race Relations 107
Ellen L. Bassuk	Homeless Children: An Emerging Tragedy 110

Robert T. Michael	The Gender Wage Gap 112
John L. Palmer	Child Care in Disarray 115
Lawrence S. Lewin and Dean R. Gerstein	Effective Drug Treatment 117
Stephen E. Fienberg and Miron L. Straf	Making Sense of Statistics in the Courtroom 120
Richard H. Thaler	The Economics of Reality 122

6

SCIENTIFIC HORIZONS

John C. Gordon	Knowing About Trees 127
Eugene H. Levy	Together to Mars — But with Deliberation 130
Peter H. Raven	The Less-Noticed Worldwide Revolution 133
Charles A. Bookman	Searching for Buried Treasure 136
David L. Morrison	The Energy Crisis Beyond the Persian Gulf 138
Herbert A. Simon	The Challenge to Human Uniqueness 141
Bruce M. Alberts	Making a Map of the Human Chromosomes 144
Luigi Mastroianni, Jr.	Developing New Contraceptive Options 146
David L. Crawford	Farewell to the Night Sky 149
Frank Press	Setting Our Science Priorities in Order 151

7

INTERNATIONAL AFFAIRS

H. Guyford Stever	Getting Even in International Technology 157

Peter W. Likins	The Growing International Competition for Brain Power 160
George R. Heaton, Jr.	Industrial Cooperation in Japan: It's Not What We Think 162
Paul C. Stern	Offering Tools for Soviet Democracy 165
Robert W. Kates	The Surprising Reality About Hunger 167
Phyllis Freeman	Vaccines for the Developing World 170
Julie DaVanzo	Easing the Fear of Giving Birth 172
Hugh Popenoe	New Crops for South America's Farmers 175

8
DIFFICULT CHOICES

Ralph Crawshaw	Life and Death: More Than an Expert Opinion 181
Dorothy Nelkin and Laurence Tancredi	The New Diagnostics and the Power of Biologic Information 183
Alexander Morgan Capron	Harvesting Organs from Anencephalic Infants 186
Ronald Bayer	HIV Screening and the Calculus of Misery 190
Norman Hackerman	Laboratory Experiments on Animals Should Continue 192
Arthur H. Rubenstein and Rosemary Chalk	Integrity and Science 195

9
THE NEXT GENERATION

Eugene E. Garcia	Kindergarten Stress 201
Philip and Phylis Morrison	Abe Lincoln's Schoolroom 203
Gilbert T. Sewall	What School Volunteers Can Do 205
Bernard L. Madison	The Challenge of Numbers 208
Francisco J. Ayala	On Darwin, Bibles and Classrooms 211
Susan Coyle	The Long Haul to a Doctorate 214
Paula Rayman	The 'Mommy Track' in Science 216
Willie Pearson, Jr.	Dr. King and Blacks in Science 219
Samuel C. Florman	The Civilized Engineer 222

INDEX
225

All of the articles and author affiliations in this book appear as originally published.

Foreword

Frank Press
President, National Academy of Sciences

We live in an era of unprecedented scientific and technical achievement. Our lives have been transformed by computers, medical breakthroughs, space probes and a host of other changes — as well as by such dangers as environmental degradation and nuclear conflict. My own field of geophysics, in which we have learned how the continents move across the globe on tectonic plates, illustrates the fantastic progress that has occurred in a wide range of scientific disciplines.

Yet, despite their importance, the many developments in science and technology remain mysterious to millions of Americans. People have little understanding of semiconductors, genetic engineering, global warming and other issues that are changing their lives. Opinion surveys and tests of U.S. students' knowledge show that public understanding of science and technology is weak. Even Americans with advanced training in non-scientific fields often know little about the revolution in biology or the amazing new materials being produced in laboratories. As science journalist Ben Patrusky writes in this book, "When it comes to science, many smart, achieving, curious and otherwise accomplished folk come down with a case of synaptic shutoff. Here we are living in what is truly an astonishing era in human history, a dazzling epoch of scientific and technological achievement — mere prelude to marvels yet undreamed of — and many of its beneficiaries are indifferent to, if not altogether bored by, the very enterprise that brought us to this most exalted station."

This popular indifference poses a challenge to a democratic society facing important decisions about AIDS, drugs, national defense, medical technology and other issues involving science and technology. How can voters and policymakers act wisely about issues they do not understand? Our economy requires excellence in science and technology, from the factory floor to research in such cutting-edge fields as optics, biotechnology and microelectronics. International competition in both basic research and technology is increasing steadily, and people's jobs and prosperity are certain to be affected by how well our nation fares. Americans also will encounter many questions involving science and technology in their daily lives, from evaluating the risks posed by radon or pesticide residues to deciding whether to purchase safety devices for their automobiles, what to eat and how to preserve the environment. In the world of the 1990s and beyond, knowledge about science and technology is not a frill but a necessity.

The path that separates scientists from non-scientists runs both ways. Many scientists, engineers and other technical experts make an inadequate effort to explain their work and concerns to their fellow citizens. Even if their specialty bears directly on important issues of the day, these experts may be reluctant to venture beyond their classroom or laboratory to speak with public officials, journalists or even at a neighborhood gathering. Many of them, trained in the rational ways of the scientific method, find the larger world maddeningly illogical and imprecise. Yet this is precisely why more of them need to roll up the sleeves of their lab coats and share their special expertise with the larger society. Former New Jersey Governor Thomas Kean, writing in this volume, is correct in arguing that, "In these days of complex problems and high-tech solutions, it is essential that those who understand the laws of nature be more involved in the making of the laws of man."

Headline News, Science Views seeks to bridge this gap between science and the rest of society. It features 75 brief essays by some of our country's most prominent scientists, engineers, physicians and other experts. The authors discuss terrorism, space travel, rising sea level, sustainable agriculture and other issues in language that is remarkably free of jargon. They outline issues in a way that makes science and technology interesting even for people who struggled to pass high school chemistry. All of the articles originally were syndicated nationally by the National Academy Op-Ed Service. The volume was edited by David Jarmul, who has directed the service since its inception in 1983. We hope *Headline News, Science Views* will help bring the worlds of scientists and non-scientists a little closer. Neither group is well-served if, like the continents, they continue to drift apart.

Washington, D.C.

Editor's Note

The articles in this book originally appeared on the editorial and opinion pages of daily newspapers. They were distributed by the National Academy Op-Ed Service. Begun in 1983 under the auspices of the National Academy of Sciences, National Academy of Engineering, Institute of Medicine and National Research Council — the institutions whose reports serve as the basis for many of the articles — the service provides more than 250 newspapers with timely articles by scientific and technical experts. The papers receive the weekly articles free with exclusive rights within their cities. Among those that have published stories from the service are *The Atlanta Constitution, The Boston Globe, The Chicago Tribune, The Cleveland Plain Dealer, The Detroit News, The Houston Chronicle, The Miami Herald, Newsday, The Philadelphia Inquirer, The San Francisco Chronicle* and *The St. Louis Post-Dispatch.*

The wonderful cartoons and drawings in this volume were published originally by editors at subscribing newspapers. The artists and editors granted us permission to reprint the illustrations here.

The service would not exist without the continued support and encouragement of the editors at the subscribing newspapers, who have helped us bring these complex scientific and technical issues into the arena of public debate.

We also are indebted to hundreds of study committee members, staff officers and others within the Academy who have shared their expertise and offered advice on story ideas. The entire staff of the Academy news office supports the Op-Ed Service in many ways, from reading drafts to gathering clips; Gail Porter, former director of the office, and Patricia Worns, the copy editor, contributed to every article in this volume.

Our greatest thanks is reserved for the authors, who took time out from busy schedules to prepare these articles without pay and under tight deadlines. Making the transition from scientific text to newspaper prose was not always easy, but it was made much smoother by authors whose prominence was matched by their patience, eloquence and genuine desire to reach out beyond the scientific community to the American public.

David Jarmul

HEADLINE NEWS, SCIENCE VIEWS

1

SCIENCE AND NON-SCIENTISTS

Getting the Facts Straight About Science

Bill Cosby

I had a lot of misconceptions about scientists when I was a kid.

Like about Thomas Edison. My teacher told us one day that the world would have been a very different place if Edison had not invented the light bulb. The electric light didn't seem like such a big deal to me, however. After all, we could still watch television by candlelight.

I got confused about Archimedes, too. He was the guy who went running naked through the streets of Athens shouting "Eureka!" after he sat in a bathtub and discovered that his body pushed out some of the water. All I could think of when I learned about old Archimedes was that if I'd spilled water all over my bathroom floor, my *mother* would have shouted and it wouldn't have been "Eureka!" I didn't even want to think about running naked through the streets of Philadelphia.

In fact, the only scientist who I really appreciated was the great Italian astronomer Galileo Galilei. I loved his name. I imagined having a first and last name myself that sounded so much alike, like Cosby Cosbei. *That* would have sounded cool on the basketball court.

What makes me remember these experiences is that May 12–18 is National Science Week, and I'm afraid that a lot of American kids have the same mistaken notions about science and scientists that I did when I was in school.

Drawing by Robert Barnum
Fort Wayne News-Sentinel, Ind.

There is an unfortunate image of scientists as old guys in white coats who putter around laboratories saying things like: "Where did I put my data?" The fact is, however, that scientists are men and women of all ages who are doing some truly fascinating things.

We all know about astronauts, of course. But scientists also are searching for cures for cancer, sickle cell anemia and other illnesses. They are improving methods for detecting air and water pollution, understanding how to deal with toxic waste dumps and spotting new environmental problems like acid rain. They are developing new energy devices like solar-powered electric cells.

Scientists also are seeking new ways to ease hunger and disease in Africa and other parts of the developing world. For example, they have begun using genetic engineering and other new techniques to develop better seeds and farming systems so Africans can grow enough food even during a drought. Biologists are working on vaccines for malaria, sleeping sickness and some of the other diseases that kill millions of Africans each year.

All this is *interesting* and it's worthwhile, too. Scientists may not wear sequined gloves or play Sunday doubleheaders, but young people should recognize that a career in science can be exciting. Scientists send rockets to study the planets and climb into tiny submarines to learn about the spreading ocean floor. They race to erupting volcanoes, fly into hurricanes and are among the first ones called when some strange disease strikes a community.

The spread of computers has turned on a lot of young Americans to the fact that science can be interesting, worthwhile and a good way to make a living. Science also provides the understanding needed to create new jobs and industries and keep our country competitive economically.

Nonetheless, many young people either have not gotten the message or else remain unconvinced. One reason is that kids are not exposed to science enough. The average American elementary school student studies science for just 25 minutes per day, and most of that time is spent reading from textbooks rather than doing experiments. Students typically learn about plant growth and other life sciences, but get little training in such interesting subjects as chemistry or physics.

The situation gets worse in the higher grades. Many of our country's high school students do not study physics, chemistry, biology or astronomy at all these days. This contrasts sharply with the situation in Japan and Western Europe, where high school students generally receive much more extensive instruction in science.

Schools across the United States have become aware of this problem and are trying to improve their instruction of math and science. But the schools cannot do it all. Parents need to show kids how exciting science is by taking them to museums, zoos, universities and farms. Airlines, factories and other businesses should invite students to see science and engineering in action.

National Science Week offers a special opportunity for these kinds of activities. Schools, businesses, community groups and other organizations across the country are sponsoring science fairs, exhibits, tours and other events to increase public understanding of science and technology. The events have been organized by the National Science Foundation, private companies, professional societies, educators and others.

Parents and students should take advantage of these events. But we also need to encourage our young people to pursue science year-round. After all, they are the ones who will have to make the new discoveries and design the future. They need to learn, as Thomas Edison might have put it, that science is fun; there's nothing that can hold a candle to it.

May 12, 1985

Bill Cosby, *star of "The Bill Cosby Show," holds a Ph.D. in education. He served on a commission that studied pre-college education for the National Science Board, the policy-making body of the National Science Foundation.*

★ ★ ★

Who Killed Yankee Ingenuity?

Steven L. Goldman

What do "Back to the Future III," "Gremlins 2: The New Batch" and "Robocop 2" have in common, besides being movie sequels? Like their own predecessors, like "Die Hard 2" and "Total Recall," and like many other recent popular films, they offer hostile or distorted images of scientists, engineers and technology.

"E.T.," for example, triggered a series of films in which scientists are depicted as heartless, or so committed to solving a technical problem as to be blind to its moral implications. Among these films were "Splash," "Iceman," "Baby," "Project X" and "The Manhattan Project." One wonders how many of the tens of millions of young viewers were inspired to pursue a career in science by the spectacle of scientists determined to dissect or otherwise abuse such lovable creatures as E.T., a beautiful mermaid, a revived neolithic man, a baby dinosaur and highly intelligent chimpanzees.

The Disney studio has a long history of depicting scientists as genial bumbling fools, and this continued with the father in "Honey, I Shrunk the Kids." But the critical edge was much sharper in the depiction of the insufferable, compulsive physicist father in "Parenthood."

Technology frequently appears in films as a means of extending corporate, military or political power regardless of its impact on people or nature. In "Robocop" and "Robocop 2," for example, corporate greed controls technological innovation. In "Tucker: The Man and His Dream," corporate manipulation of a U.S. senator ruins Tucker for having developed a superior automobile. In "Total Recall," corporate control of the planet Mars is an amoral tyranny sanctioned by the government on Earth because of the military applications of the alloys the corporation manufactures.

In "The Mosquito Coast," the engineer hero is a neurotically self-centered genius who wants to impose "straight lines and right angles" on a "curved" Nature. His technical triumphs prove temporary, but the damage he inflicts on his wife and children, on forest and river, is lasting. Similarly, the heros of "The Conversation" and "The China Syndrome" are engineers who, for a long time, deny moral responsibility for the applications of their expertise. And both men turn out to be fatally inadequate to correcting the wrongs they suddenly discover they are abetting.

In "The Emerald Forest," technology — in the form of an American-built dam complex on the Amazon River — unbalances the relations between two primitive jungle tribes. The crisis that develops is resolved by destroying technology, thereby restoring the natural balance.

In the "Star Wars" trilogy — which, like "E.T.," played to huge audiences around the world — it is the Evil Empire that possesses the superior technology. The rebels win because they are morally worthy of tapping into the spiritual power of the Force.

Technological wizardry is ultimately irrelevant in "Batman" and in all of the James Bond films, too. Bond and Bruce Wayne triumph because of their heart, strength, courage and commitment to the right. In the two "Die Hard" films, the hero repeatedly expresses disdain for technology, while the whole point of "The Terminator" was the ability of a naked human being to defeat the ultimate assassin: an intelligent, virtually indestructible robot.

How could a culture that for so long has prided itself on its technical ingenuity, and has for more than 40 years supported science and technology with public funds on a massive scale, so patently enjoy seeing science and technology depicted so negatively?

More to the point, if the United States is to maintain, let alone improve, its standard of living, it must improve its industrial competitiveness. Doing so will require a greater role for science and engineering in corporate and governmental decision making, as well as more scientists and engineers in industry and academe. Where are these new professionals to come from? What is to motivate students to make the course choices in high school that will keep open the door to careers in science and engineering?

No one expects movies to promote the educational, or the vocational, needs of society. Movies are entertainment. But the themes to which large audiences respond by their ticket-buying and video-renting decisions are symptomatic of prevailing values. Unless these change, audiences will continue to jeer evil technologists while our technical expertise, and with it our prosperity, fades further into our national history.

August 5, 1990

Steven L. Goldman *is the Andrew W. Mellon Distinguished Professor in the Humanities at Lehigh University.*

* * *

On an Antidote for Science Phobia

Ben Patrusky

I had a revelation recently as I was watching a pro football game. At one point the announcer went about diagramming a just-completed play. It was a real doozy and I had a hard time following what all those X's, O's and arrows were up to. But, trying my damnedest to keep up, I did, along with millions of other fans.

As I looked at the diagrammed play on the electronic chalkboard, it came to me suddenly that all those X's and O's looked very much like the pictures produced when nuclear particles are made to collide in atom smashers.

Now imagine if TV viewers were somehow suddenly confronted with a blow-by-blow analysis of particle collisions in accelerators. The collective channel clicks would register eight or higher on the Richter scale.

When it comes to science, many smart, achieving, curious and otherwise accomplished folk come down with a case of synaptic shutoff. Here we are living in what is truly an astonishing era in human history, a dazzling epoch of scientific and technological achievement — mere prelude to marvels yet undreamed of — and many of its beneficiaries are indifferent to, if not altogether bored by, the very enterprise that brought us to this most exalted station.

Confront a science phobic and he or she may argue, "Absolutely not true." Some will swear that they adore science. After all, they'll say, they devour their newspaper's weekly science section. They never miss Carl Sagan on the tube. My argument is that it's not science they're reveling in, but the products of science — new technologies, new cures, new answers to life's woes and enigmas.

Most of these folks, I would contend, show little curiosity about science itself — about what scientists do, about how they conduct their investigations. The results are great, but the arduous journeys required to get there seem to be of little interest.

This widespread aversion to science comes from Americans' not knowing the complete story about how science actually happens. In football we know a lot about the rules of the game — the process — and also about the players' lives: their salaries, their off-field exploits, their drug habits.

That's hardly the case with science. By and large what the public has gotten is a mythic view of science along with a scientist-as-demigod iconography. The emphasis is on the products of the magic box, not on the process that leads to the creation of those products. Put another way: The public gets little of the X's and O's and arrows and very few three-dimensional glimpses of the players. It gets merely the scores. And that, inevitably, breeds alienation, incomprehension and fear.

As a science writer I discovered rather late in my career how flawed this perception really is. As I spent more and more time in the laboratory watching scientists do their thing I discovered that not all of them were geniuses. Some were mediocre or just plain inept. Many, even the best of them, were often wrong. As Newton is my witness, I have seen scientists manifest foolishness, arrogance, jealousy, vengefulness, ambition, envy, egotism, territoriality — and, heaven forfend, even out-and-out dishonesty.

Human qualities don't just disappear with the donning of a lab coat, but they seldom surface in descriptions of contemporary science. Oh, once in a rare while, events force us to get a glimpse of them. Witness the story of human failings — and courage — that came to light in the wake of the Challenger tragedy. However, under normal conditions the human dimension is usually absent.

None of this is meant to diminish the glory that is science. On the contrary, I would contend that if the human element were a more normal part of the telling about science, then science itself would prove far more tantalizing and attractive and its practitioners much less priestly and remote.

From whence sprang this mythic perception of science and scientists? The major perpetrators, I think, are our schools, which teach science in a way that treats learning and fun as being mutually exclusive. Also, some scientists have been content with keeping the public in a mystified thrall, and those of us who report on science have tended to cover scientists with "gee-whiz" awe and kid-gloves respect. This only sustains the myths that foster fear amongst the populace.

Those of us who care about science need to share with the rest of the public our well-kept secret that it is very much a human endeavor, practiced by flesh-and-blood folk. By humanizing science we can ease our country's science phobia.

November 22, 1988

Ben Patrusky, *a science writer and communications consultant, is executive director of the Council for the Advancement of Science Writing. This article is adapted from a longer version that appeared in* Issues in Science and Technology.

★ ★ ★

Physics for Poets, Science for Society

Leon M. Lederman

National surveys assure us that public understanding of science has never been poorer while our national need for a scientifically literate public has never been greater. We are faced with an endless array of issues — AIDS, pesticides, ecological Armageddon, space stations and Stealth bombers — that are inextricably entwined with science and technology. How are we to manage these issues if we do not understand them?

Scientists and science communicators must do more to enhance "science literacy" among their fellow Americans through books, television programs, museums and, perhaps most productively, our nation's schools and colleges.

I recently retired as director of the Fermi National Accelerator Laboratory in Illinois. Fermilab is a dynamic laboratory, operating the world's most powerful particle accelerator. My years there were fruitful and, about a year ago, I was offered a third five-year term as director. I could have looked forward to a scientific bonanza from our new superconducting accelerator, the Tevatron. It took us ten years to build this machine, and now I could let young experimenters develop the physics results while I basked in a reasonable glow of reflected glory.

Instead, as I looked around at the problems our country faces, I became convinced that there is a call as challenging and perhaps more pressing than the cutting edge of physics research. I accepted an offer to teach a "Physics for Poets" class at the University of Chicago, which I began this semester. Physics for Poets is an alternative to other courses, such as "Rocks for Jocks" in geology, that liberal arts students can select to satisfy their science requirement.

My decision was personal, but I do think it illustrates the deep concern that many scientists feel about the importance of enhancing the capacity of the American public to deal effectively with the threat of planetary environmental catastrophe and other problems involving science and technology.

It is simply unacceptable that so many students come out of college as innocent of science as when they entered. Our society must have citizens comfortable enough with science to participate in decisions that will ensure our safe passage into the 21st century. By the same token, many of our students who *do* study science need a richer appreciation of the social sciences and humanities. They must do better in communications and be more aware of the ethical dimensions of their work.

This gap must be bridged and I want to help. There is a selfish component to my decision, since the only way for a person of my generation to hang onto a remnant of the clarity of perception characteristic of the young is to seek out young people and teach them. Yet I also know how important it is for future politicians, newspaper editors, judges and voting citizens to feel comfortable with scientific concepts.

Before coming to Fermilab, I was a student and professor at Columbia University for 32 years. The chairman of the physics department there, the great I.I. Rabi, insisted on strong teaching as well as outstanding research. I began my own teaching with "Physics for Nurses" and evolved to teach general physics for science students and advanced courses for physics majors.

Finally, I got the job of teaching physics to liberal arts students in the same kind of class that I am now teaching at the University of Chicago. The students were intellectually lively; one equation would make their eyes glaze. I had to prepare each lecture meticulously, asking myself what I wanted them to recall in their later lives.

Today, I am convinced that we will not have effective political decisions unless our leaders and the voting public can better cope with scientific and technological concepts. The university is one good place to start. There is nothing that compares to the hot action of seeking simplicity and order in the sub-nuclear jungle, but I look forward to contributing to the Herculean task of forming a new academic coalition to redefine the requirements of a 21st century college graduate.

A scientifically educated citizenry and a concerned scientific community cannot remain just a desirable goal for our country. Increasingly, as we face technological accidents of global scope, the hole in the ozone layer, the terrifying global warming trend and so many other issues, it is becoming the price of our collective survival.

October 15, 1989

Leon M. Lederman, *Director Emeritus of Fermilab, is a member of the National Academy of Sciences. He was co-winner of the 1988 Nobel Prize in Physics.*

★ ★ ★

Making Sense of a Risk-Filled World

John Ahearne

The Alaskan oil spill has faded from the headlines and flowed to our subconscious as further proof of the perils of modern life. It resides there beside Chernobyl, Bhopal, the Challenger, Three Mile Island and other disasters.

Our world often appears to be a dangerous place. We are bombarded with warnings about chemical residues on our foods, radon in our basements and dozens of other hazards. Yet we also enjoy longer and healthier lives than our grandparents. So how is one to interpret all of these alarms and decide where action really is needed?

That is a dilemma not only for average citizens who hear about risks in their daily lives, but also for the scientists, engineers and public officials who must help the public assess the risks associated with everything from nuclear reactors to food additives.

An expert committee of the National Research Council, which I chaired, recently found that many of these technologists are frustrated in communicating their views on risk. Go to their conventions and you will find sessions on "Explaining [topic of the meeting] to the public." Listen and you will hear them discuss in baffled tones why so many of their fellow Americans get enraged over some threats that are statistically small while remaining nonchalant about other dangers that are demonstrably — to technologists — more serious.

Americans on both sides of the technical fence have a problem with "risk communication," and it is due largely to several widely held misconceptions.

The first of these is that disputes over risk are always about facts. Many technologists believe the public would agree with them if only it would stop complaining and get educated. These experts view opposition to their recommendations as proof that people are not getting the message, whether because of inattention, the complexity of the subject or some other reason.

The problem with this view is that many disputes are not about facts at all. They are about values. People simply may disagree with an industry or public agency about the relative importance of jobs, energy, safety or a pristine environment. Arguing over the details in such a dispute, as so often occurs, is unlikely to resolve these differences.

A related misconception is that improved communication alone can eliminate conflicts like these. Wrong again. Although better communication certainly is desirable, it may cause people to solidify their position if it clarifies the differences in their underlying values.

Still another misconception — and one of the most widely held — is that a major part of the problem is journalists who sensationalize events to sell papers and boost ratings. This sometimes does occur, but journalists are essentially intermediaries, transmitting the messages of others. They cannot be blamed when those messages contain bad news or are poorly constructed.

Both technologists and the public need to stop thinking of risk communication as a process in which learned experts issue information to a placid audience. Instead, it must become a two-way street in which all sides genuinely interact. That is easier said than done, of course; any parent, child, employer or employee knows that real listening is difficult. Yet only through such exchanges can people get beyond the flood of numbers to the underlying values.

To make this happen, the public should demand better performance by government and industry leaders. When these leaders meet with reporters to discuss a potential hazard, they have a responsibility to present the facts — and the uncertainties — in a way people can understand. Clear communication cannot be regarded as a frill to the "real business" of technical competence. It is not only the public that needs complete and understandable information, but presidents, governors and other senior officials who may lack technical training.

The public also has a right to honesty. Government officials are servants of the people, and they should tell the truth without sugar-coating or putting a "spin" on it. Honesty is the prerequisite not just to credibility, but also to making decisions on the basis of facts and public values rather than on ideology.

It serves no one well for technologists to complain endlessly about the public's lack of technical training or for average citizens to feel they are being treated in a condescending manner. Experts and non-experts must work together if we Americans are to begin making better sense of our technological society.

November 12, 1989

John Ahearne *is executive director of the scientific research society Sigma Xi.*

★ ★ ★

Making the Link Between Science and Politics

Thomas H. Kean

My experience as governor of New Jersey has convinced me of a growing split in our country between policy decisions and rational scientific thought at a time when the problems facing us have become overwhelming in their complexity.

Whether the subject is pollution, health care or industrial revitalization, we politicians need the help of the scientific community more than ever before. Our problem is not a lack of information — we are drowning in data. What we need is for scientists to overcome their traditional aversion to politics and help us make decisions more wisely.

Too often, the "experts" who now come forward are pseudoscientists. They are otherwise intelligent people who believe their Ph.D.'s in English or education make them PDQ: pretty darned qualified to explain the subtleties of science to the public. They are dentists and accountants who hold up their "advanced degrees" as they call for an immediate stop to this or increased funding for that, all in the name of science. They play to the six-second sound bite, and they do it well: Their brand of "yellow science" is remarkably effective in energizing people and motivating politicians.

The result is often what the commissioner of the Food and Drug Administration, Frank Young, has called "risk assessment by media." This is no way to decide public policy. If we politicians are to separate the serious from the spurious, we must have competent scientific help.

Unfortunately, too many scientists consider elected officials to be lower forms of life — missing links in the evolutionary chain. These scientists prefer to work on their own worthwhile research rather than wade into the political mud. Of course, many scientists do share their expertise with policymakers, but their numbers are too few to meet our needs.

I am not advocating the Morton Downey-ization of science. I don't expect to see Nobel laureates on daytime television hectoring Geraldo or opening up to Oprah. But scientists must make themselves more available to the public. They should attend fewer faculty meetings and more town meetings. They must appear not only on *Nova*, but on the nightly news. An educated public is the secret to our common survival. Science itself has a term for those who are too old or too slow to learn: extinct.

As they become more involved, scientists must take care not to scare the public with dire predictions based on time frames that are short for scientists but long for the general public. Warnings of Saharas in the Dakotas or beachfront property in Dubuque lose their impact if they fail to happen quickly. California has yet to slip off into the Pacific, and that has put public trust in scientific prediction on shaky ground. Scientists must always keep their audience in mind and give it specific, realistic recommendations.

Beyond helping public officials and reaching out to the

public, more scientists should consider leaving the halls of science to work in the halls of Congress. This great industrial society is represented in its federal and state legislatures by hundreds of lawyers — but only a handful of chemists, biologists, engineers and others with technical training. America needs people with scientific expertise to throw their hats — and their lab coats — into the political ring to help make sound decisions about the environment, space policy, agriculture and other critical subjects.

By raising their level of involvement in these ways, scientists can help rid society of important misconceptions about what society can and cannot expect science to accomplish. In the past, science has spoiled us with its success. Time and again, it has found the silver bullet that can cure whatever ills society faces. But we cannot continue to create messes, counting on science to clean up after us.

Times have changed, and we must change with them. It is no longer acceptable for a governor to duck a tough decision by saying, "I'm a politician, not a scientist." For the good of our children, governors must know about both politics and science — and scientists must, too. In these days of complex problems and high-tech solutions, it is essential that those who understand the laws of nature be more involved in the making of the laws of man.

July 30, 1989

Thomas H. Kean *is governor of New Jersey. This article is adapted from a speech he gave at the National Academy of Sciences.*

★ ★ ★

TECHNOLOGY IN EVERYDAY LIFE

Making Our School Buses Safer

Charley V. Wootan

School is opening this week and millions of children will board buses without safety belts. That may come as a shock to many parents who have diligently required seat belt use in their own cars. Now they must allow their children to ride unrestrained in a much larger vehicle.

It doesn't seem to make sense. Here our schools are determined to protect our kids from everything from AIDS to asbestos, yet most school buses in the United States lack safety belts. Wouldn't it be a better use of our resources to buy and install them?

Surprisingly, it might not.

I recently chaired a committee of the National Research Council that studied this question in detail, and we discovered — contrary to what one might expect — that there are more effective ways to protect young bus riders than with seat belts. In fact, we concluded that the overall potential benefit of requiring seat belts in school buses is insufficient to justify any new federal standard, although states and local school systems might want to install them on their own initiative.

In an average year, 10 children die while riding in large school buses — which, unlike smaller buses, generally do not have seat belts. By contrast, nearly 40 children are killed annually while trying to board or leave a bus. About two-thirds of those 40 are struck by a school bus, usually their own.

Drawing by Ned Levine
Copyright, 1989, Los Angeles Times Syndicate. Reprinted by permission.

Such a relatively low death rate is impressive when one considers that the nation's 390,000 school buses travel nearly 4 billion miles annually. Riding in a school bus is four times safer than riding in a passenger car.

Still, 40 or 50 deaths of children cause acute suffering for the families involved, and we ought to reduce this toll if we can. Our committee, which included experts in highway safety, pediatrics, bus manufacture, occupant-restraint systems and other fields, calculated that equipping all school buses with seat belts would save one life and avoid several dozen serious injuries each year. The cost would be more than $40 million per year.

Alternatively, it would cost only $6 million annually to raise the height of seat backs from the standard 20 inches to 24 inches. This would save up to two to three lives and up to 95 injuries a year.

In terms of both cost and lives saved, in other words, raising seat backs is probably more effective than installing seat belts. It also avoids the problem of requiring drivers to keep their eyes on the road while ensuring that all the children actually use the belts. Any parent who has tried to do the same with lively children in the back seat knows this is easier said than done. It *is* reasonable, however, to expect bus drivers to prohibit their passengers from standing in the aisles while the bus is in motion. More states should require this, which means ensuring enough seats for everyone.

More generally, states and school systems ought to focus on where most fatalities actually occur, which is along roadsides and in school loading zones. Too many children are killed when they dart in front of buses or play near them. Many of these deaths could be prevented with better safety training for both children and drivers and by installing stop-signal arms and improved cross-view mirrors on buses. Other possible techniques include loudspeakers to warn children, crossing-control arms to guide them as they cross the street, and electronic or mechanical sensors to help bus drivers detect children in blind spots outside the buses. All these warrant further study.

One of the best ways to improve safety is by removing from operation as quickly as possible buses manufactured before April 1, 1977. These older buses are not as strong structurally and have less padding on the seats, lower seat backs and less "compartmentalization" to contain children in the event of an accident. They also are more prone to post-crash fires, as occurred in the terrible 1988 crash of an older bus in Carrollton, Ky., which killed 27 occupants.

Raising the height of seat backs, prohibiting standees, improving bus driver and student education, testing new safety devices and getting old buses off the road all would reduce deaths and injuries significantly without burdening already strained school budgets. Seat belts are valuable, too. Yet, to save as many children as possible, we ought to concentrate our efforts where they will do the most good.

September 3, 1989

Charley V. Wootan *is director of the Texas Transportation Institute at Texas A&M University.*

★ ★ ★

A Computer Future Without a Heart

Samuel H. Fuller and Damian M. Saccocio

Like the Tin Man in "The Wizard of Oz," the United States faces a future without a heart. For all the euphoria we Americans may feel about recent political changes around the world, we are threatened by a less-publicized economic threat to the heart of one of our most vital industries and, thus, to our collective prosperity.

The industry is computers, and our country's position in it is far more precarious than most Americans realize. After all, the computers and software that most people use in their offices or homes are made by U.S.-based companies. It certainly appears that, unlike the situation with video cassette recorders or automobiles, computers are one industry where our country remains on top.

But appearances are deceptive. The U.S. computer industry is actually a collection of several connected industries — from component manufacturers to software designers. Together with telecommunications, they account for a tenth of our country's gross national product.

All of these diverse enterprises depend on the microchip, the incredibly small and powerful device that is the fundamental building block for all computers. Microchips make possible the word processors, computerized banking, video games and many other innovations that have revolutionized our lives.

A National Research Council report released this past week suggests that the microchip heart of our computer future is deteriorating, with potentially grave consequences not only for the industry but for U.S. society generally. Production of microchips and semiconductor equipment was pioneered in the United States. Yet, during the past five years, one U.S. company after another has withdrawn from the marketplace in the face of intense foreign competition, particularly from Japan.

As happened with many consumer electronic goods,

A COMPUTER FUTURE WITHOUT A HEART

Drawing by Bill Hogan
The Record, Hackensack, N.J.

Americans excelled at developing the first microchips, but Asian firms came on strong in producing them in large quantities efficiently and with high quality. Asian manufacturers now produce over 90 percent of the world's dynamic random access memory (DRAM) chips. Unlike most U.S. firms, they endured a glut of these essential chips in the mid-1980s and are now reaping the profits from a $10 billion market.

If current trends continue, U.S. computer hardware manufacturers may find themselves limited to specialty markets, which is roughly comparable to our automobile industry agreeing 50 years ago to produce only jeeps and convertibles. This new situation threatens a loss of jobs and profits not only for U.S. chip producers themselves, but also for companies in related industries. Asian firms already have claimed about 45 percent of the U.S. market for personal computers and, together with European firms, are making inroads in the software arena.

Of course, cheap personal computers and other products are a boon to American consumers, but not at the cost of our semiconductor industry. Our national future depends on computers and information systems, and we simply cannot write off our core industry in this field. It is essential that U.S. chip manufacturers fight back by becoming more global and strategic, pursuing new technology and opportunities worldwide. They must not only invest in long-term research, but also do a better job of making the less-glamorous incremental improvements that often spell the difference between success and failure.

They cannot compete alone; they need the assistance of other U.S. companies and researchers, as well as of government. Our traditional cult of the solo entrepreneur has limited applicability to this task, given how closely foreign megafirms work with their governments and bankers. In fact, cooperation among our own companies, universities and government has contributed to many of our country's successes in technologies ranging from computer networks to parallel computing. We need more such collaboration.

Our computer industry is now a justifiable source of national pride, but a failure to recognize its precarious position risks taking us down a path already traveled by U.S. manufacturers of automobiles, cameras, steel and other products. The challenge posed by foreign companies and governments is formidable, and we must become more strategic and decisive in meeting it. We cannot allow our heart to just wither away.

January 28, 1990

Samuel H. Fuller, *vice president of research at Digital Equipment Corp., chaired a colloquium on the future of the U.S. computer industry for the Computer Science and Technology Board of the National Research Council.* **Damian M. Saccocio** *is a staff officer with the board.*

* * *

Toward Motoring Smart

Robert D. Ervin and Kan Chen

As millions of commuters well know, traffic in most metropolitan areas is straining the highway system to its capacity. If present trends continue unchecked it will get even worse. Fifteen-minute delays suffered by the typical commuter in 1988 may stretch to an hour by early in the next century. Congestion will prevail throughout the day. Accidents will become even more disruptive.

The traditional solution to traffic congestion has been to build more roads. But we are reaching the point where we can no longer build our way out of the congestion crisis. Land and money are too scarce and many communities oppose new highways. What is needed now is truly radical change, including a more imaginative application of technology to vehicles and the highway system.

For example: A common bottleneck on many highways is toll-collection booths. These could be eliminated with a system that assigns each vehicle a code number and then scans them as they pass by, like items at a supermarket. The system would send each vehicle owner a monthly bill.

Roadside transmitters could send messages to voice synthesizers inside vehicles, informing drivers about road conditions. Vehicles could be equipped with compact discs that provide electronic maps of any neighborhood in the country. They might have radar or laser systems to warn them of collisions with other vehicles. Technology built into the highway might regulate the position of all vehicles, controlling merging and exiting.

Such concepts of "smart transportation" have received a fair amount of publicity lately, but too many of the stories fail to look beyond the "gee whiz" nature of the technology to how we actually put it in place.

We must learn from Europe and Japan, which have been exploring these options aggressively through joint public- and private-sector efforts. The most comprehensive European effort involves 20 automotive manufacturers and 70 research institutes from six countries. Non-European parties are excluded. Japan, likewise, has been going it alone in developing intelligent transportation concepts with an eye toward eventual commercial markets.

Research on "intelligent highways" in the United States has been much more modest and dispersed, although lately there have been some hopeful stirrings. Cooperative research programs have been established at the University of California, Berkeley; the University of Michigan; and Texas A&M University, College Station.

The three domestic U.S. automakers and various electronics firms are starting to participate in long-range research on intelligent highways. The U.S. Department of Transportation has provided some support.

Still, a much greater research effort is needed to address the many questions that remain unanswered. How, for example, will transportation planners handle the transition to this new system, when some vehicles are equipped with new features while most are not? From a legal standpoint, who will shoulder the blame if an accident is caused by a glitch in the automated roadway: the driver, the highway department or the car maker?

Until now technological and social concerns like these have kept both the private and the public sectors in our country in a state of inaction. U.S. companies have shuddered to invest in technology whose horizon for new products is years away. Government transportation agencies have been unwilling to undertake research without assurances that it will solve real highway problems.

To escape this chicken-and-egg situation we need more than tinkering with conventional approaches; we need vision. The Big Three car companies, in particular, must provide leadership and resources to make the concept credible. Highway agencies, too, must take a longer-term perspective. The only way to develop this new paradigm of a vehicle-highway system is for vehicle designers and road builders to work together. Doing so will also require a commitment from political leaders and a strong federal role because the states are unable to conduct the research and field trials of these systems by themselves.

We have no alternatives as a nation but to proceed vigorously with a joint government-industry effort. The magnitude of the U.S. traffic problem requires that we innovate in dealing with our vehicle-highway operations, providing American industry with a setting in which to work with government in developing the high-tech option. Otherwise, the collaborations already under way overseas will enable Europe and Japan to claim this new marketplace uncontested. It is time we also steered into the future.

February 21, 1989

Robert D. Ervin *is a research scientist at the Transportation Research Institute at the University of Michigan.* **Kan Chen** *is a professor of electrical engineering and computer science at the University of Michigan. This article is adapted from a longer version in* Issues in Science and Technology.

★ ★ ★

Easing the Crunch at Our Airports

Joseph M. Sussman

The volume of air traffic is rapidly outpacing the capacity of our nation's airport system. The current level of 1.3 million passengers daily is expected to double by early in the next century.

That could mean trouble for anyone who flies.

The boom in air travel, spurred by deregulation and other factors, has strained our major airports. In 1987, 21 of them experienced at least 20,000 hours of airline flight delays. The need for additional and better utilized facilities is acute. Yet only two new airports have opened during the past 20 years, one near Dallas, the other at Fort Myers, Florida.

Just one major airport is now under construction, outside Denver. Possible new airports in Los Angeles, Austin and northwestern New Mexico will not be operational before 2000, if they are built at all. Airports under consideration in Atlanta, Chicago, Minneapolis-St. Paul, San Diego and St. Louis lie even further in the future.

In a report being released today, a committee of the National Research Council said the outlook is bleak unless something is done to accommodate the continuing growth of air travel. Air gridlock would be agonizing for travelers and would harm the economy.

Adding airport capacity through new runways, terminals and parking lots, as well as with entirely new airports, is one alternative. Yet complex factors involving noise, environmental damage, budget limitations and other problems that must be considered often prevent the addition of flights or the expansion of existing airports, much less the building of new ones. Given the slow pace of current expansion and the likelihood of further political opposition to new airports, we also must think creatively about other options.

One possibility is to shift some traffic to airports that are now underused, creating new "airline hubs," as has been done by Delta at Salt Lake City and Orlando and by American at Raleigh-Durham and Nashville. This would ease the

Drawing by David Wink
The Atlanta Journal-Constitution, Ga.

burden on overloaded hub airports elsewhere and reduce delay throughout the system.

An even more innovative, although unproven, idea is to develop airports specifically designed as transfer points. More than half of the passengers who arrive at Chicago's O'Hare and Atlanta's Hartsfield, for example, are headed somewhere else. Why not have these passengers change planes at a less crowded location and reserve major metropolitan airports for people who are actually traveling to these cities?

Managing demand, through either administrative centralized management of the air network or such market measures as peak-hour pricing for airport access, can lead to better use of existing capacity — albeit not without controversy.

New technology also can help. Aircraft equipped with quieter engines may be able to operate less obtrusively, enabling airports to increase operations or to stay open longer without bothering nearby residents. Improvements in the air traffic control system will allow runways to be used more effectively. A new generation of widebody jets, meanwhile, may be able to carry between 700 and 1,000 passengers, reducing the number of flights needed to serve densely traveled routes.

New aircraft that can operate on very short runways or that take off and land vertically could replace conventional planes for shorter flights while requiring much less space at airports. This could free longer runways for large jets or allow air service at smaller satellite airports near major population centers.

Another possibility is to develop high-speed surface transportation to replace air service in heavy travel corridors. Passengers who now fly between, say, New York and Boston might be attracted to a ground-based system using high-speed rail or magnetic levitation, or to "smart" highways that allow smart vehicles to travel with computer assistance.

There is no single, simple answer. All these and other ideas require extensive study and testing and a much more vigorous research and development effort. The federal government, working in cooperation with state and local authorities and with the private sector, must move decisively to avoid saturation of airports and airways. Action is needed now to lay the foundation for a safe, convenient and affordable intercity travel system to carry us into the next century. Our current system simply cannot stretch to twice its size without breaking.

October 14, 1990

Joseph M. Sussman, *director of the Center for Transportation Studies at Massachusetts Institute of Technology, chaired a National Research Council committee that studied the capacity of U.S. airports.*

★ ★ ★

Protecting Our Phones from Terrorism

John C. McDonald

A growing terrorist threat in the United States is as close as your nearest telephone. Although deregulation and the introduction of new technology have improved our national phone system in many ways, they also have left it more vulnerable than ever before to terrorism and other perils.

That might not appear to be as alarming as a terrorist attack on other targets, such as an airliner or a municipal water supply, but the situation threatens far more than Sunday chats with Grandma. Major disruption of telephone lines could prevent air traffic controllers from communicating, cut off police and fire services and wreak havoc on businesses that send large amounts of data over the lines. Our information society could be brought to a screeching halt in the blink of an eye — or the cutting of a cable.

A glimpse of this perilous future was provided a year ago when fire damaged a large telephone switching facility near Chicago. Tens of thousands of persons were without the emergency 911 service for many days. Some businesses suffered large financial losses and O'Hare Airport shut down temporarily.

The U.S. phone system is increasingly vulnerable not only to terrorism and fires, but also to computer hackers, natural disasters and inadvertent damage from construction digging. Internationally, most of the phone calls that connect us with the rest of the world will soon be carried by a small number of fiber-optic cables whose location on the ocean floor is well-known. These, too, are possible targets for terrorists.

The situation has worsened following the breakup of AT&T, which operated a national center that was responsible for maintaining phone service during emergencies. Today, that responsibility has shifted to a government agency, the National Communications System. The NCS is professional and hard-working, but the task it faces is ever more difficult.

The NCS must deal with several companies instead of just one, and many of these companies — such as AT&T, MCI and Sprint — have different technological standards, equipment and rules. Instead of securing a single network, in other words, NCS has to cope with a "network of networks." In today's competitive environment, furthermore, these companies cannot easily invest in costly security measures for which there is no immediate payback. In the past, AT&T provided security and emergency measures as a normal cost of doing business.

Beyond these institutional problems is the double-edged sword of technology, which has made long-distance calls not only clearer and cheaper, but also more vulnerable. Advances in fiber-optics technology, for example, now make it theoretically possible for the equivalent of all the telephone calls at a given moment in the United States to fit into just one cable. Although lines of this capacity are not yet in service, the trend clearly is towards concentrating more and more calls onto fewer fibers.

Digital switching technology is concentrating the network in a similar fashion, routing far more calls than did previous devices. Over the next few years, the 19,000 central switching offices in the United States will give way to a much smaller number of offices with switches that are far more powerful. The computer software that lies at the heart of this emerging system, meanwhile, also has grown more concentrated.

Moreover, this software will become available outside the phone companies — and hence to any determined terrorist — to allow enhanced-service providers to compete with the local-exchange carriers on an equal technical footing.

To assure reliability, the National Security Council should address this growing vulnerability of the U.S. telephone system and consider steps to reduce it. For instance, security might be improved at critical network nodes and more diversity and redundancy might be built into the system. Crisis-management capabilities could be improved, such as by requiring companies to provide priority service to police, hospitals and other selected users during declared emergencies. More should be done to protect telephone software from hostile use.

As matters now stand, a few well-placed grenades could bring down large portions of our domestic long-distance networks, cutting off calls from Hartford to Honolulu. We need to protect the system, now, before that happens.

August 6, 1989

John C. McDonald, *executive vice president for technology at Contel Corp., chaired a National Research Council committee that studied the growing vulnerability of the U.S. phone system.*

★ ★ ★

The New Arsenal of Democracy

Robert B. Kurtz

It's hard to remember that before Iraq's forces rolled into Kuwait, Americans were debating how to spend the "peace dividend" produced by the end of the Cold War. Now those hopes seem naive. All of us have been reminded that the world remains a dangerous place.

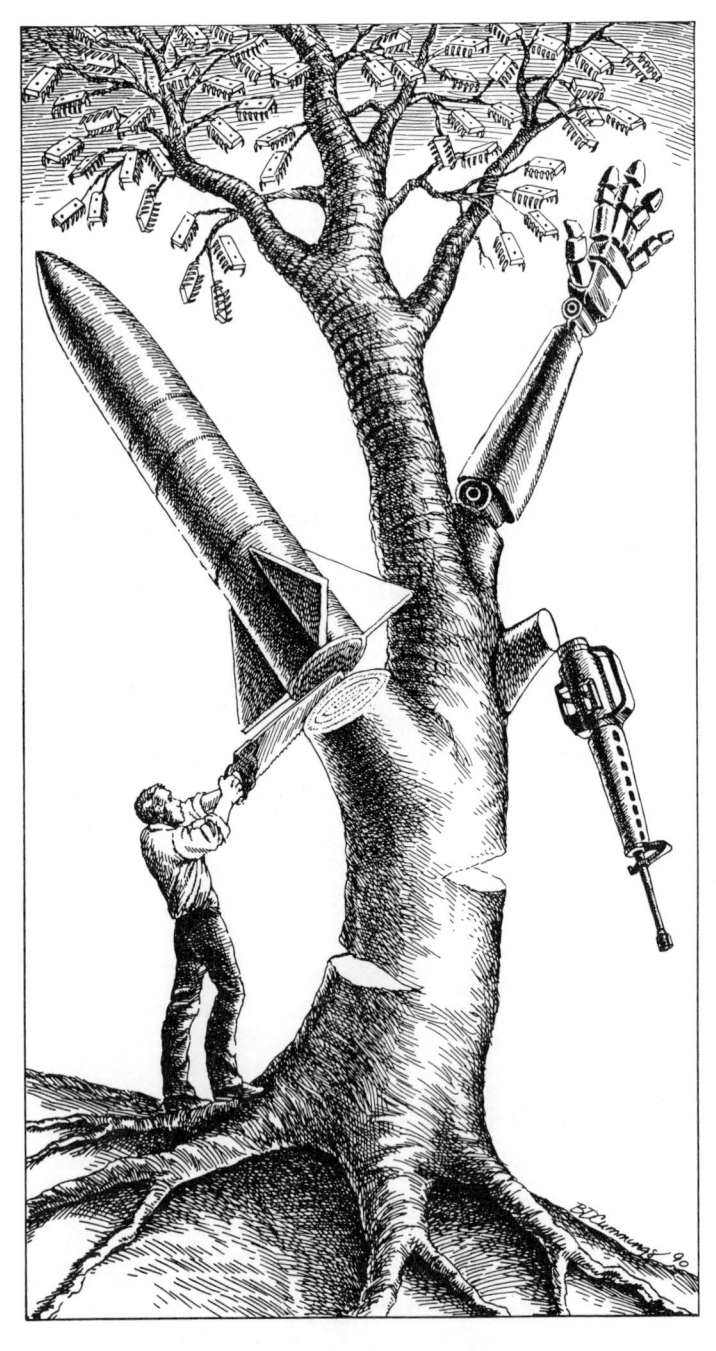

Drawing by Barbara Cummings
Copyright, 1990, Los Angeles Times Syndicate. Reprinted by permission.

Saddam Hussein has illustrated in the boldest colors that the United States cannot beat swords into plowshares without having a backup plan to convert them back into swords. Although we may not need the same weapons or troop levels we had during the height of the Cold War, we must retain the capacity to meet new crises.

This includes rapid deployment forces, fighter planes, ships and other military assets. But something else must be on our list as well, something less obvious but perhaps even more important: the industrial capability to produce weapons and supplies quickly.

That capability is diminishing at a disturbing rate. I chaired a National Research Council committee that examined this recently and we came away worried about how well, and how fast, the United States will be able to respond to conventional wars in the future.

In today's technological world, it will be much harder to gear up for an extended fight than it was during World War II, when "Rosie the Riveter" and millions of other Americans worked around the clock to produce the guns, tanks and other supplies for the nation's "arsenal of democracy." Today, Rosie might need a degree in computer science and the skills to work on an advanced fighter aircraft, not to mention the right equipment, materials and support system.

Many of these resources are increasingly uncertain in the United States. Suppose, for example, that a modern Rosie was asked to design a microchip for an aircraft guidance system. She might find it difficult, if not impossible, to locate a domestic company with the equipment to convert her design into an actual product. U.S. firms that produce such critical products as bearings, machine tools and computer components have declined severely in the face of intense international competition. U.S. companies also are battling in such fields as optics, sensors, ceramics and superconductors, all of which have potential military applications.

A single component produced by a U.S. firm can be essential to a complex weapon system. If the firm goes bust, the Pentagon must go elsewhere. If every other U.S. company also drops out of the market, as has occurred in many consumer electronic fields, the Pentagon must buy the component overseas. That leaves the weapon system, and in turn our defense, vulnerable in a world in which we must expect the unexpected. Just imagine if we couldn't get our hands on the computer chips needed to run our planes, submarines and tanks.

U.S. defense contractors should be modernizing their plants to avoid this possibility. But, even before recent cutbacks within the industry, defense firms were investing inadequately in manufacturing technology. Many companies lack the sophisticated machinery needed to produce modern weapons. One reason is that the Defense Department provides little incentive to them to operate more efficiently. More attention is given to designing weapons than to ensuring that they can be manufactured.

Many companies also lack excess equipment for emergency defense production. In an era of corporate raiders and intense competition, they can no longer afford to keep equipment idle waiting to be revved up during a crisis. They are poorly prepared to cope with the inevitable bottlenecks that a sudden gear-up would cause.

The problem is not only within the companies, but at the Pentagon. The Defense Department is aware of the need for industrial preparedness, but its process for ensuring it is fragmented and inadequate. President Bush should create a national group to set clearer policies for industrial preparedness and Defense Secretary Cheney should elevate responsibility for the issue to a higher level. Other top planners also must give the problem more attention.

The gap between our military needs and our industrial capability is widening and in danger of becoming unbridgeable. Beyond the horizon of the Persian Gulf, this eroding capacity to fight conventional wars poses a real threat to our security. There will always be another Saddam, Khaddafi or Noriega, and we need to be prepared for them.

August 26, 1990

Robert B. Kurtz *is a retired senior vice president of the General Electric Co.*

★ ★ ★

Building Houses People Can Afford

Ezra Ehrenkrantz

Why is housing so expensive when televisions, computers and many other goods cost relatively less now than they did before?

Between 1970 and 1987, the consumer price index tripled, but the price of clothing and telephone services only doubled. The median sales price of a privately owned one-family house, on the other hand, jumped from $23,400 to $104,500 — a breathtaking increase of 447 percent.

Breathtaking and, for people trying to purchase their first home, heart-stopping. Only 20 percent of Americans now earn enough to purchase a new house at market rates without a trade-in, a dramatic drop from 50 percent two decades ago. For millions of people, the dream of owning a home has faded.

The main difference between houses and televisions, of course, is that houses require land, which is in fixed supply with rapidly escalating costs. Housing prices also are affected by interest rates, local business conditions and other factors that are hard to ameliorate.

But one factor that can and should be changed is the outdated way we build houses. Modern building techniques could reduce the cost of a new home from, say, $100,000 to $90,000, or even less. That is not a huge difference, but every dollar counts, particularly when one computes interest costs over the life of a mortgage.

Most builders in our country now produce houses one by one with conventional materials instead of taking advantage of mass production techniques and newer technology. They install bathrooms one fixture at a time rather than using prefabricated units with the lights, toilet, sink and tub already in place. They do the same for kitchens and make inadequate use of breakthroughs in composite materials, microelectronics and robotics.

The lowly two-by-four remains the primary construction material, even though a growing demand for wood products has caused it to become scarcer and more expensive. Few

American home builders have thought seriously about replacing two-by-fours, a sharp contrast with the situation overseas, where many builders are experimenting actively with alternative materials and systems.

The failure of the construction industry to innovate threatens its own future in the same way that technological complacency hurt U.S. automobile and steel manufacturers. In some states, segments of the construction industry are now dominated by foreign companies.

For frustrated home buyers who lack the money even for modest "starter" homes, the situation is already critical. It will probably get worse so long as housing follows the characteristics of a service industry rather than a manufacturing industry. The aging of the baby boom generation and other trends may provide some relief, but low productivity will keep many Americans in rental units instead of their own homes. Those at the bottom of the economic ladder, in particular, will face rising rents and fewer viable options.

Home builders are not inherently averse to new technology. However, there now is little incentive within the industry to invest in technological innovations. Developing new technologies is expensive, requiring not only basic research but also material testing, construction of prototypes, code approval, tooling for production and marketing of the final products. Any one of these activities may take several years.

As things stand, the would-be innovator has no way of knowing what interest rates, the money supply and other conditions essential to success in the housing market will be like when the product is finally ready. As a result, over the past 15 years the building industry has tended to make minor changes to existing products rather than invest in true innovations.

For the sake of millions of would-be home buyers, this needs to change. One of the best ways the industry could become more innovative is through new public and private programs that spur fresh concepts and new products. Test beds should be established to try out appropriate ideas, facilitate testing and speed regulatory approval of innovations. To succeed, experimental programs would need to protect prototype designs and a limited number of housing units from frivolous lawsuits, and to disseminate their results widely.

Instead of wringing our hands endlessly about housing costs, it's time we tackled each of the components of that cost and, with respect to technology, became more creative about supporting research and development. Americans need houses they can afford.

December 31, 1989

Ezra Ehrenkrantz *is president of Ehrenkrantz, Eckstut and Whitelaw, an architectural firm in New York. He holds the Chair of Architecture and Building Science at the New Jersey Institute of Technology and is a member of the Building Research Board of the National Research Council.*

★ ★ ★

Designing for an Aging America

Sara J. Czaja

Despite the fact that the elderly are the fastest-growing segment of the U.S. population, everyday environments often make life difficult for them. For example:

- Labels on medicine bottles often are small and difficult to read.
- Water faucets, door knobs, and lids on jars and other containers can be hard to open or close.
- Kitchen counters and cabinets sometimes are too high to reach, especially for people with arthritis.
- Appliances often are difficult to operate.

Most of these and hundreds of similar problems are nuisances, but some are life-threatening. I recently co-chaired a National Research Council committee that found that many of the problems could be prevented with forethought and better design. The four examples cited above, for instance, might be remedied with larger medicine labels, new faucets, lower kitchen counters and less-complex appliances.

During the 1990s, the number of people in the United States aged 55 and over will increase by 11.5 percent, a gain of more than 6 million persons. The growth rate among those aged 75 and older will be 26.2 percent, or a gain of nearly 4.5 million people.

Many of our homes, workplaces and highways were designed for a younger population. This is problematic; people aged 65 and over account for approximately 43 percent of all home fatalities. Many need help carrying out routine activities. Yet, in general, their special needs have received remarkably little study. Research is lacking on how declines in physical and mental agility that occur with aging affect the ability to carry out routine activities.

For example, while it is known that the strength of a person's grip tends to decline with age, the implications for the design of kitchen appliances have been studied insufficiently. Similarly, we do not understand how declines in vision, memory and reaction time affect one's ability to drive an automobile.

This is not to suggest that the needs of the elderly have been ignored; many companies produce excellent products for older customers. Neither is aging synonymous with frailty or illness. Many older Americans enjoy healthy, productive lives. But increased research on the "human factors" needs of the elderly would be of enormous value. Wherever possible, the goal should be to create designs not specifically for the elderly, but for the entire population. Safer bathtubs or stairs, for example, would benefit everyone.

High priority should be given to living environments. Approximately 13 percent of older people who live at home exhibit at least one major decline in physical mobility. Elderly people also are more likely to live in older homes that need maintenance and repair. Many end up in nursing homes for lack of relatively simple changes in their homes. That is a tragedy for those involved and costly for society.

Similarly, an unknown number of older workers retire every year because their companies are inflexible in meeting their specific needs. Older employees may require extra lighting or a bit more time to perform certain tasks, but they more than compensate in other ways, such as with their greater experience. More research also should be done by transportation planners on everything from how to redesign roadways to providing easier baggage handling in airports.

Technology can solve many, although not all, problems. Scalding accidents in bathrooms are common among the elderly, for example, and they could be reduced by redesigning water controls, lowering water pressure or regulating the temperature. New remote monitoring devices can track the vital signs of elderly people or the movements of Alzheimer's patients who may wander.

One inspiring application of technology is occurring in the Miami area, where a group of 38 elderly women are using home computers to send each other messages and view the latest news, weather and movie reviews. The women, the oldest of whom is 95, already have used the system to produce a cookbook. They now are anxious to learn new applications, such as banking or shopping from home.

Society needs more innovations like this for its homes, offices, roads and shopping centers. After all, none of us is getting younger. Our "built environment" ought to age gracefully along with us to meet our changing needs.

September 23, 1990

Sara J. Czaja *is research director at the Stein Gerontological Institute in Miami and associate professor of industrial engineering at the University of Miami.*

★ ★ ★

Preparing for the Next Big Natural Disaster

Richard E. Hallgren

The dust has settled from the Bay Area earthquake, but what about the next big quake? What if it strikes not in California, which is relatively well-prepared, but in some other part of the country whose susceptibility to earthquakes is less well-known?

Charleston, for example, is still digging out from Hurricane Hugo, but a major earthquake occurred there in 1886, killing 60 persons. Missouri has a high probability of experiencing an earthquake registering six points or greater on the Richter scale within the near future, and 80 percent of the building stock in St. Louis is unreinforced masonry, which could crumble in a quake.

In fact, a panel of the National Research Council cautioned earlier this year that Boston, Seattle, Memphis and other urban areas in at least 39 states face significant risk.

The contrast between what occurred in the Bay Area and the earthquake of similar magnitude in Soviet Armenia last year illustrates the danger that many of these other U.S. cities face. Most would experience far fewer than the estimated 25,000 deaths that occurred in Armenia, because building codes and construction techniques are different from those in the Soviet Union. Nevertheless, without adequate preparation, the devastation from a major quake in many U.S cities could be severe.

The problem is not only earthquakes, but natural disasters generally. Hurricane Hugo, Hurricane Jerry and the Bay Area earthquake followed on the heels of wildfires in the West, flash floods and debris flows in Hawaii and an earthquake in southern California during the past two years. Nine hundred tornadoes strike our country each year, and more than $1 billion in damage is caused annually by landslides.

Although we cannot prevent these disasters from occurring, we could be doing much more as a nation to prepare for them and minimize their impact on life and property.

A group of leading scientists, engineers and others reported recently to the Research Council that the United States is failing to take full advantage of a growing body of knowledge about hazards. Our country's current approach can be characterized as a patchwork of temporary fixes, incomplete analyses of alternatives and uncoordinated actions and policies. In engineering designs, investment decisions, and public and private policymaking, many efforts — such as building on landfill in San Francisco or on barrier islands in South Carolina in the face of known hazards — increase potential losses rather than reduce them. Proven technologies are not applied, whether out of a false sense of economy or inertia. The United States also spends much less than many other countries on research and development in this field.

Certainly, many improvements have been implemented. One look at the survival of San Francisco's skyline or at the early warning system in Charleston shows that substantial progress has been made and that many dedicated Americans are working hard to save lives. Yet the same two disasters also revealed that much more remains to be accomplished.

Ironically, the earthquake and hurricane struck at a time when the United States has a unique opportunity to expand its efforts in hazard mitigation, both at home and internationally. This past week, the United Nations acted to implement a worldwide program that designates the 1990s as the "International Decade for Natural Disaster Reduction." The program will encourage scientists, engineers, public officials, urban planners and others from around the world to share research and information on hazard-reduction techniques.

For Americans, this need not require losing perspective about the tradeoffs involved in cost, land use, convenience and the like. Much can be gained from relatively simple measures, such as encouraging homeowners in storm-prone regions to secure their roofs, controlling erosion to prevent landslides and preparing better relief plans. More expensive measures, such as reinforcing older highway structures like the one that collapsed in Oakland, may also be worth the cost.

The most important need, however, is to overcome complacency about the situation. Both Hurricane Hugo and the

California earthquake show that Americans must not be fatalistic about disasters. We must grab the historic opportunity before us to make these events much less disastrous.

November 2, 1989

Richard E. Hallgren, *executive director of the American Meteorological Society, chairs the U.S. National Committee for the Decade for Natural Disaster Reduction.*

* * *

Our $1.5 Trillion Investment

Robert F. Jortberg

Imagine a national asset whose value dwarfs the cost of the Clean Air Bill, child care, fighting drugs and other initiatives combined.

Such an asset exists, but it has the misfortune of falling under the heading of "infrastructure." I would wager that most people who read that word immediately think of bridges or potholes. Either that, or they turn to the comics page.

Infrastructure involves more than transportation, however, and it is not mundane. A significant piece of it is as close to our daily lives as our neighborhood school, hospital and post office. These and other public buildings are worth approximately $1.5 trillion. Many of them are quietly falling apart.

The parking garage atop the New Haven Coliseum corroded so much that it now must be demolished and rebuilt after less than 20 years. One urban school district has a maintenance budget so low that its classrooms will be painted only once every 100 years. Buildings at hospitals, universities and other public institutions are saddled with everything from leaky roofs to severe structural problems.

A family severely in debt would not allow its car to operate without oil because it cannot afford a new one. Our country is in a similar position with its severe budget deficit, yet it is failing to maintain and repair its buildings. Department of Defense buildings alone are worth more than $500 billion, and the nation's 88,021 public school buildings are close behind.

I chaired a National Research Council committee that recently examined this situation, and we were troubled by what we found. Many officials are failing to protect public buildings, and the potential costs of correcting the situation already total billions of dollars. Poorly maintained buildings not only threaten the health and safety of occupants; they also are terribly demoralizing for the teachers and others who must work amid broken lights and heating systems.

The problem is not that officials deliberately sabotage schools, prisons, fire stations, recreation centers and other buildings. Most of them are hard-working people trying to serve a public that wants more services without new taxes. Yet when faced with a choice between maintenance or providing more visible services such as snow removal or police protection, many officials choose the latter.

Similarly, it is inevitable that politicians will reap more glory for breaking ground at a new building than for modernizing the ventilation system at an existing one. Only if that ventilation system becomes a breeding ground for Legionnaire's disease, or if some other calamity occurs, will they suffer. More likely, the consequences will become apparent only after the politician leaves office.

These failings are understandable, but the fact remains that public officials are supposed to be stewards of public assets. In cases where political expediency motivates the decision, neglect of building maintenance is nothing less than a squandering of public funds.

What can be done to improve the situation? The answer does not lie in castigating public officials but in reforming the process that leads them to make the decisions we now see on the federal, state and local levels. Our committee offered two specific suggestions.

First, officials at all levels of government need more and better information. Too many city managers and state public works directors are trying to allocate funds without a clear understanding of what's needed. At a minimum, they should assess all buildings regularly in a way that yields useful information. Techniques exist to accomplish this, and officials ought to start using them.

Second, budgets should reflect the fact that adequate maintenance and repair are essential parts of the overall cost of owning public buildings. An amount equal to about 2 percent to 4 percent of buildings' replacement value needs to be dedicated to this function each year. Officials should not be allowed to "borrow" these funds for other purposes.

The larger need, of course, is money. Increased spending on maintenance would do wonders and, in the long run, actually save funds by eliminating the need for many repair or replacement projects. These are our buildings, all $1.5 trillion worth of them. If we fail to protect them, we'll just end up spending more to rebuild them.

July 15, 1990

Robert F. Jortberg *is a retired Navy rear admiral.*

* * *

Tough Choices About Rising Sea Level

Robert G. Dean

The world's sea level is slowly rising. Does this mean you should avoid investing in a beach condominium? Should coastal communities be planning to retreat from the oceans? Should they, instead, try holding off the waters with dikes, storm gates and pumps? Or is all this a lot of paranoia over nothing?

One thing is sure: Rising sea level and coastal deterioration are on the minds of a lot of Americans. *Time* magazine recently published a cover story entitled "Shrinking Shores" that was filled with grim tales of families in coastal states watching their dream homes sink into the ocean. Communities from Maine to Maui are worried about the problem.

An expert committee of the National Research Council, which I chaired, recently tried to set the facts straight about this growing issue. We released a comprehensive study of what sea level rise means for coastal planners and engineers. Our committee considered a range of possible increases in sea level over the next century — from about 20 inches to nearly five feet — and pondered their environmental and policy implications.

Our basic conclusion was that there is no reason for panic. Boston, Miami and San Francisco are not about to slip into the ocean. On the other hand, the evidence is convincing that sea level *is* rising and that many coastal communities lack the data they need to make realistic decisions about coping with the situation in the decades ahead.

Most scientists agree that increased levels of carbon dioxide in the atmosphere, caused in part by our modern appetite for fossil fuels and forest products, are creating a "greenhouse effect" that will raise world temperatures by three to eight degrees Farenheit over the next century. Scientists disagree, however, over the extent to which this will cause the world's glaciers to melt and ocean water to expand, bringing about a rise in sea level.

Changes in sea level are measured in two ways: the relative change between the land and the sea in a given location and the absolute change worldwide. During the past century, relative sea level has risen by about 12 inches along the Atlantic coast, six inches along the Gulf of Mexico and four inches along the Pacific coast. Much of this increase has been due to land subsidence, or sinking, caused by the extraction of underground water or hydrocarbons, by earthquakes and soil compaction, and by other factors. In the Louisiana delta, for example, these factors and wetlands destruction are responsible for the sea's moving tens of feet inland every year.

At the same time, relative sea level has been *falling* in Alaska and other far northern locations as glaciers melt and the land is relieved of their tremendous weight.

In general, most shoreline communities in the United States *do* need to start getting ready to deal with a higher sea level over the next century. They have three options: armor the shoreline, nourish it with dredged sand or retreat to higher ground.

None of the options is cheap. Armoring a shoreline with sea walls and other devices or nourishing it with dredged sand can cost as much as $800 per foot of shore front. In the case of beach resorts or port facilities, of course, the existing investment may be so great that these expenditures are warranted. Elsewhere, retreat may be the best choice. North Carolina, with its environmentally fragile barrier islands, for example, has already established an erosion buffer zone where new construction is prohibited. Other states are doing the same.

Fortunately, most beach cottages and other small buildings with short life expectancies are not threatened immediately by rising sea level — although the land on which they are built may eventually wash away. Similarly, most industrial facilities are renovated often enough that rising sea level can be accommodated. Here, too, it depends on local conditions and other factors.

A greater problem is the fate of coastal bridges, airport runways, power plants and other major long-term facilities. These expensive installations may have to be protected from rising waters or redesigned to handle the new conditions. Also of special concern are the intrusion of salt water into underground aquifers and the impact of rising sea level on coastlines that have special environmental, as opposed to economic, value.

Again, it is important to stress that global sea level is rising slowly enough to prepare for its effects. But the situation clearly demands increased vigilance and research rather than complacency. Specifically, much more intensive monitoring and research programs are needed to help coastal communities make these multi-million-dollar decisions more knowledgeably. The time to start getting ready is now, before the waters rise any higher.

October 11, 1987

Robert G. Dean *is a professor in the department of coastal and oceanographic engineering at the University of Florida, Gainesville.*

★ ★ ★

3

A SUSTAINABLE FUTURE

Uncertainty and the Greenhouse Effect

Robert M. White

Global average temperatures have broken records during the past five years. U.S. farmers in the Midwest have suffered severe droughts. The amount of carbon dioxide in the atmosphere continues to increase. Does all this prove that the "greenhouse effect" has begun, bringing us a warmer Earth, rising sea level and a host of unpleasant consequences?

Before you rush to say "yes," consider that the same question might reasonably have been asked after the "Dust Bowl" of the 1930s — years that were followed by a period of cooler global temperatures.

Global warming has become an international scientific and political happening in recent months. Yet how robust is the scientific basis for this outpouring of worldwide concern? Do recent increases in temperature truly signal a permanent trend — or are they just a fluctuation in the perennial cycle of drought and flood?

The situation is not as certain as some headlines would have us believe. Yet, rather than being an excuse to do nothing, this uncertainty challenges us to choose our policies with even greater shrewdness and care.

The concentration of carbon dioxide and greenhouse gases in Earth's atmosphere has increased steadily in this century as a result of the burning of fossil fuels, the production of chlorofluorocarbons (CFCs), worldwide deforestation and other forces. By the middle of the next century, this concentration is likely to be twice as great as it was at the beginning of the industrial age.

Mathematical computer models of the oceans and atmosphere project that, in response, the planet's temperature will rise at least several degrees Fahrenheit by the middle of the next century. These models are only approximations of the real atmosphere, but their results, while uncertain, must be taken seriously.

The climatic response judged by actual temperature readings from around the world during the past century is less conclusive. Global average temperatures reveal a roller coaster course during the past century with a small net increase of only about 1 degree Fahrenheit. Some temperature records — for example, for the United States — reveal no evidence of change.

Most estimates of the social and economic consequences of global warming are based on scenarios of the future course of climate change and models of economic development. These scenarios can only provide information about the range of possibilities and not predictions of how climate changes will affect agriculture, water resources or ecological systems.

In other words, despite all the headlines, what we have is an inverted pyramid of knowledge, a growing mass of proposals for action balanced upon a handful of real facts.

What policy directions, then, make sense? At the very least, we need to invest immediately in improving the information base. We also should adopt policies that will ease the situation without foreclosing our future options if projections turn out to be incorrect.

Many of these policies involve the production, distribution and end use of energy. We must shift away from coal and oil to natural gas where feasible, increase efficiency in energy production and end uses, and develop passively safe and publicly acceptable nuclear power, as well as other non-fossil energy sources. All such actions will reduce emissions of carbon dioxide.

Energy-policy actions have great leverage because they often help solve other environmental issues, such as acid rain and local air pollution, while reducing our dependence on foreign oil. However, they are inherently controversial because they imply economic burdens, raising the question of who pays. Any policy framework also must recognize the international nature of the issue and its potential divisiveness between developed and developing nations. Why, for example, should Brazil protect its Amazon forest or China reduce its reliance on coal? To be serious about international action, a global bargain is necessary.

If we cannot arrest the processes of climate change, then we will need to adapt to them. We should think in terms of a continuous policy process where we periodically reassess our responses in light of new findings. Highlighting the uncertainty surrounding climate change in this way runs the risk of being interpreted as a delaying tactic. That would be the worst policy of all. The task before us is to try to step wisely even though our path remains obscured by an uncertain atmosphere.

August 27, 1989

Robert M. White, *president of the National Academy of Engineering, was the first administrator of the National Oceanic and Atmospheric Administration.*

★ ★ ★

Saving Sea Turtles

John J. Magnuson

Long before there were Teenage Mutant Ninja Turtles, our nation's southern coastal waters abounded with the real thing. Sea turtles are among our most distinctive animals. Some grow as big as 1,000 pounds.

These evolutionarily ancient reptiles appeared on earth eons before humans. But now, five species of sea turtles that live in coastal waters of the Atlantic Ocean and the Gulf of Mexico are threatened by human activity — in particular, by shrimp trawls.

The Kemp's ridley turtle faces the most serious danger. As recently as 1947, about 40,000 females came ashore on one day at the species' only nesting beach. Currently, an estimated 350 mature females nest *each year* at the same Mexican beach along the Gulf of Mexico.

The other four species also are officially threatened or endangered. I chaired a National Research Council committee that has just completed a detailed study of the decline and conservation of these turtles. We concluded that shrimp trawls are killing more sea turtles than all other human activities combined.

Turtles also may be hit by boats, ingest plastic debris, lose their beach habitats, be caught for sport or food, or die in other ways as the result of human activity. But shrimp trawls are responsible for the most deaths, by far. These nets are dragged along the ocean bottom to capture shrimp. The number of turtles caught dead or comatose in the trawls increases dramatically after about 50 minutes of towing because turtles must surface at least once an hour to breathe. Most fishing crews keep their trawls in the water for more than an hour at a time for economic reasons.

The numbers of dead turtles found stranded on beaches increase when shrimp fisheries open in South Carolina and Texas and decrease when a shrimp fishery closes in Texas. Similarly, nesting populations of loggerhead turtles are declining in Georgia and South Carolina, where shrimp fishing is intense. Yet nesting populations are stable or, perhaps, increasing farther down the coast in southeastern Florida, where shrimp fishing is rare or absent.

Previous studies estimated that shrimp trawls kill about 11,000 sea turtles annually. However, after reviewing studies of turtle mortality, counts of stranded carcasses and other evidence, our committee concluded that the actual total may be as much as three or four times higher. A disproportionate number of the animals killed are of prime age to contribute to the breeding populations.

A solution exists to reduce this toll. Turtle excluder devices, or TEDs, can be attached in the nets to allow almost all turtles to escape. TEDs are trap doors that release turtles and other large marine animals from a trawl. Some shrimpers complain that TEDs also allow many shrimp to escape, and their argument has some merit. Although certain TED designs retain almost all of the shrimp under good conditions, performance may suffer from seaweed or debris on the bottom.

Still, the devices are the best method available to conserve populations of the five species — the Kemp's ridley, loggerhead, green, hawksbill and leatherback sea turtles. To conserve these species successfully, TEDs must be used in shrimp trawls at most places from Cape Hatteras to the Mexican border and during most times of the year.

In addition, a greater effort must be made to protect beaches where turtles lay eggs. Many of these critical nesting areas are disturbed by pedestrians, recreational vehicles, sand-cleaning equipment and housing developments. A related problem is artificial light from street lamps and beachside homes; lights can disorient hatchlings during their journey toward the ocean. Several communities have already adopted "lights out" ordinances for the period when turtle eggs are hatching; others should follow their lead. Communities also should consider establishing reserves on important nesting beaches.

"Headstarting" sea turtles by removing eggs from natural nests and hatching the young in captivity for later release is a useful research tool — but it cannot substitute for more essential conservation measures.

Much remains to be learned about the behavior, ecology and physiology of sea turtles. However, their greatest source of danger is now clear — and preventable. The shrimping industry must begin using TEDs routinely, or these timeless creatures will face extinction.

May 27, 1990

John J. Magnuson *is a professor of zoology at the University of Wisconsin.*

★ ★ ★

Who Owns Antarctica?

Hugh Downs

A few years back, when the U.S. polar research vessel *Hero* lay at anchor off the Antarctic Peninsula receiving visitors, someone asked the ship's colorful skipper, "Captain Lenie, who owns Antarctica?" The Dutch-born Pieter Lenie hesitated only a moment before replying, "I own it. . . . It belongs to me."

The anecdote, recounted by author Michael Parfit in his book *South Light*, reflects the proprietary interest in Antarctica felt by those few people privileged to visit, work or even live for a time on the Earth's coldest, driest, most desolate and forbidding continent. It carries also a tinge of the annoyance these people feel toward the interloper.

Antarctica in recent years has become a tourist attraction. Drawn by the same quest for adventure, discovery and personal conquest that inspired the great polar explorers, contemporary adventurers are spending vast sums to cruise to the world's southernmost shores or fly to the South Pole.

Even some of the most vocal proponents of preserving Antarctica as a wildlife refuge have themselves visited the frozen continent and now, as advocates of leaving it alone, seem to want to pull up the drawbridge behind them. I have been to Antarctica as a member of the press and cannot imagine denying another human being access to the range of experiences to be had there: the glare of the six-month summer sun off the polar ice cap, that first eyeball-to-eyeball encounter with a curious Emperor penguin along the ice edge, the wail of a Weddell seal nurturing her pup, the sharpness of the clearest, coldest air on earth. I own Antarctica, too.

The prospect of tourists loose on the perilous Antarctic landscape is a source of dismay for the scientists who have made Antarctica their life's work. Can a continent that spans 5.4 million square miles accommodate both groups?

Science in Antarctica dates back to the earliest explorations and continues to be the cornerstone of cooperation among nations with vastly differing ideologies.

In the last decade, Antarctica has evolved from a land locked in ice and mystery into a living laboratory of global processes. The unfolding story of the ozone "hole," discovered over Antarctica, is an excellent case in point. Observations in Antarctica showed that man-made chemicals — CFCs — cause the depletion of stratospheric ozone, since observed on a smaller scale globally.

Studies of the West Antarctic Ice Sheet reveal the extent of the so-called "greenhouse effect" and resultant global warming, which would raise sea level worldwide. And Antarctica is an observatory, too. The high polar plateau offers a unique vantage point for ground-based observations of the upper atmosphere and beyond.

Government officials who manage American activities in Antarctica are concerned about the safety of visitors, their impact on the fragile antarctic environment and the potential for disruption of the research effort.

Antarctica is a very dangerous place. The wind comes up, sudden and fierce. The clouds descend, and particles of light dance back and forth between them and the ice surface, creating the disorienting condition of "white-out." The ice is lined with deep crevasses.

Yet the antarctic environment is as delicate as it is treacherous. Lacking land predators, the animals are trusting and approachable. But they are not fearless. A horde of humans traipsing through a penguin rookery incites chaos, causing brooding males to abandon eggs and separating chicks from their parents. In the extreme cold and dryness, ordinary litter does not decay.

Tourism detracts from the antarctic scientific enterprise in less subtle ways, too. However unwittingly, private expeditions to Antarctica put themselves in the hands of the governments conducting research there. To rescue people imperiled by weather or an accident requires the diversion of aircraft and crews — at great personal risk — dedicated to the support of science. Visitors wishing to call at research stations and take guided tours interrupt people for whom optimum working conditions exist only a few months each year.

Yet tourists have a right to experience Antarctica, too.

The icy continent beckons, inspires and challenges the human spirit.

However, if we stumble in, heavy-footed and heedless, we risk destroying the essence of its allure. Antarctica commands our attention; it demands our respect.

Thirty-seven nations have chosen to adhere to the Antarctic Treaty and convene regularly to deliberate the fate of Antarctica. The issue of tourism should be at the top of their agenda.

Antarctica belongs to everyone.

October 18, 1988

Hugh Downs *is host of the ABC News television program "20/20."*

★ ★ ★

Genetically Engineered Organisms: Monsters or Miracles?

Nina Fedoroff

The first outdoor test of a genetically engineered microorganism took place earlier this year. Gene-altered organisms are being brought out of the laboratory and into the test plot. They are almost ready for the farmer to use.

Such outdoor testing and use is called the "deliberate release" or "planned introduction" of genetically engineered organisms into the environment. The words have something of an ominous sound. Are scientists really making new life forms? Are they dangerous? Will they cause environmental problems?

To address some of these thorny questions, the Council of the National Academy of Sciences convened a special committee of biologists, of which I was a member. Its task was to examine the most frequently voiced concerns about genetically engineered organisms in the light of accumulated knowledge and experience. The results are summarized in a short publication that has just been released.

First of all, it is important to put recent developments in what is known as recombinant DNA research into historical perspective. Genetic engineering with recombinant DNA is essentially a new approach to an age-old activity, namely breeding. Human beings have been in the breeding business for centuries, producing everything from race horses and hybrid corn to magnificent roses. Breeders don't call themselves genetic engineers, but they do the same sorts of things and have the similar goals of changing organisms in beneficial ways.

Yet the new recombinant DNA techniques are powerful and special because they make it possible to move genes between organisms that don't mate in nature. An insect gene can be put into a plant, for example. But the worry that this will produce new and dangerous life forms can already be answered from experience.

Hundreds of research laboratories around the world have been doing gene transfers for more than a decade. Untold numbers of organisms with foreign genes have been produced and studied — and they have proved to be no more dangerous than their unengineered siblings. An organism is not a dangerous new life form just because it has a new gene from an unrelated organism. In fact, it's much more like one of Burpee's new improved varieties of garden vegetable.

But might these new techniques inadvertently turn an ordinary plant into a superweed that would spread like the kudzu vine that plagues the South? Could recombinant DNA accidentally convert a harmless microbe into a harmful one, like the Dutch elm disease fungus that is killing our magnificent elms?

The answer is that being a weed or a disease-causing organism is very complicated. It takes many genes of just the right kind to be a successful pest. One new gene (or even ten) won't do the job. The bottom line is that these are largely groundless worries. The genetic engineer can't turn corn into kudzu any more than a corn breeder can.

What are the concerns, then? They are the same kinds of concerns that we now have about using organisms. Some organisms, like cabbage and corn plants, aren't much of a problem at all, and we have plenty of experience in agriculture to guide us. Others require more care, depending on the nature of the organism and how it has been altered.

It is especially important to be cautious in releasing organisms into different ecosystems from those in which they evolved. Experiences with gypsy moths and Japanese beetles, to use familiar examples, have alerted us to what can happen when the interaction between an organism and a new ecosystem is not considered carefully. The scientific community needs to provide guidance in evaluating proposed introductions from an ecological perspective.

And what is the promise of recombinant DNA technology? In my view, the real promise of the new genetic engineering techniques is that they can help us solve some of our most serious problems in better ways than we could before.

To cite just one of many examples, we now kill insect pests with tons of toxic pesticides, inadvertently killing other organisms, disrupting ecosystems and contaminating the soil, our drinking water, even ourselves. Recombinant DNA techniques have begun to show us new and better ways to make pest-free crops. Scientists have recently discovered that genes can be introduced to make plants resistant to certain insects, much as we immunize people against disease.

We are faced with many pressing environmental problems. Recombinant DNA techniques can help us find ecologically sound solutions. But the endeavor is young and vulnerable. We must take care to not stifle it with excessive regulation, just as we must use the genetic engineer's new tools with skill and wisdom.

If we manage to strike a good balance between regulation and innovation, I believe we can look forward to decreasing our reliance on toxic chemicals in both industry and agriculture, and perhaps to cleaning up the world we leave our children. That's miracle enough for me.

August 30, 1987

Nina Fedoroff *is a molecular biologist at the Carnegie Institution of Washington.*

★ ★ ★

Rethinking Radioactive Waste Disposal

Frank L. Parker

Every day the amount of high-level radioactive waste at civilian reactor sites in the United States grows. By the end of this decade, the inventory will produce 1 billion curies of radiation, which is greater than the radiation released by 1,000 tons of radium.

That is a lot of waste, and some of it will remain radioactive for more than 10,000 years — a period longer than all of recorded history. Disposing of the waste safely is one of the most important environmental responsibilities we face. Yet the plan our country has adopted to do so is unrealistic; it fosters acrimony and delay instead of solving the problem.

Current plans are to build a deep geological repository for high-level radioactive waste in Nevada if the site proves acceptable, with the facility opening in the year 2010.

There is a strong worldwide consensus among scientists that this kind of geological isolation is the best and safest long-term option for dealing with the wastes, which now are stored primarily at reactor sites around the country. Yet the U.S. program is unique among those of all nations in its insistence on defining in advance the schedule and technical requirements for every part of the multi-barrier system.

This approach sounds thorough and prudent but actually is a formula for disappointment. Building a waste repository is like creating a mine in which the "ore" is put into the ground rather than taken out. Mining is a fundamentally exploratory activity; surprise is part of the underground landscape. No scientist or engineer can anticipate all of the potential problems that might arise as one explores the underground formations. It is like asking Columbus to describe the New World before he's even left Spain.

Nonetheless, given the controversial nature of the project, planners have felt compelled to try to design everything perfectly at the outset, rather than agreeing to make changes as geological features are encountered and scientific understanding develops.

One need not be "pro-nuke" or "anti-nuke" to be concerned about this. Civilian reactors have produced tons of high-level radioactive waste. The alternative to placing these in one or more permanent repositories is to continue storing them at the reactor sites. On-site storage appears to be safe for at least another century, but it may be irresponsible in the long run because of the difficulty in securing, protecting and maintaining so many above-ground sites. Even if every reactor were shut down tomorrow, the wastes would remain.

One possibility might be to move the waste to a repository that would not be closed permanently until further studies are carried out. This would allow the waste to be removed, if necessary, and would still be better than leaving them indefinitely at the reactor sites.

In any case, the United States should stop pursuing an ever-receding mirage of infallibility and adopt a more flexible, experimental policy. This does not mean proceeding recklessly, without regard for future generations. Commercial mining and underground construction both operate on the sound principle of "design (and improve the design) as you go." Canada and Sweden have adopted flexible strategies for disposing of their nuclear waste, and their efforts are much more likely to succeed than ours.

Acting without total certainty is not foolhardy; it is real life. All the facts indicate that it is feasible to place high-level radioactive waste into carefully selected underground facilities where the local geology and ground water conditions ensure isolation of the waste for many millenia, and where waste materials will migrate very slowly if they ever do come into contact with the ground water.

The public should not be placated with absolute guarantees. Most citizens watching the human frailties of their governments and technologists know better. The time has come for Congress, the regulatory agencies and the Department of Energy to adopt a more realistic approach. Continued insistence on a totally predictable system for an unpredictable world assures that our country will encounter delays, rising costs, frustration and a loss of public confidence. It also virtually guarantees that most of the nation's high-level radioactive wastes will remain where they are rather than being moved to a safer, more permanent repository.

September 2, 1990

Frank L. Parker, *professor of environmental engineering at Vanderbilt University in Nashville, chairs the Board on Radioactive Waste Management of the National Research Council.*

* * *

Toward a Sustainable Agriculture

John Pesek

One does not have to live on a farm to know that agriculture in the United States faces problems. Many Americans whose closest contact with farming is the neighborhood supermarket worry about pesticide residues in their food, water pollution due to farming, soil erosion, antibiotics in animal products or just the size of their weekly grocery bill.

Many farmers share these concerns, and a growing number of them are interested in experimenting with alternative systems that require lesser amounts of pesticides, fertilizers, antibiotics and fuel. A committee of the National Research Council, on which I served, concluded this week that these systems — often regarded skeptically by many farmers — can be productive and profitable. Farmers around the country with widely varying crops and acreage have discovered that the different methods can be as good for the balance sheet as they are for the environment.

The growing demand for safer food and a cleaner environment suggests the time is ripe for approaches like these to move into the mainstream of American agriculture, just as commercial fertilizers, synthetic pesticides, high-yielding seeds and improved machinery gained popularity after World War II. This new shift should sustain adequate food supplies well into the next century while improving environmental quality.

The new systems vary but generally emphasize diversification rather than continuous planting of fields to single or only a few crops. Diversified systems tend to be more stable and resilient, offering a hedge against the several disasters that can strike single crops. Diversification also makes it easier for farmers to reduce reliance on external inputs by rotating fields and sequencing crops in ways that enhance soil fertility, control erosion and limit pest populations.

Alternative systems, as contrasted to conventional practices, deliberately take advantage of beneficial interactions that occur naturally among crops, animals, insects and microorganisms. They emphasize management instead of purchased inputs. Computers, advances in biotechnology and other breakthroughs make it possible for farmers to do this with increasing sophistication.

Why, then, do not more farmers use alternate practices?

One reason is that the practices require more information, trained labor, time and management skills than do conventional techniques. It also is difficult for farmers to calculate the financial impact of making a change. There are many variables involved, particularly during the transition to some alternative methods. Someone with a family to feed may hesitate before changing.

A dominant reason is that federal agricultural policies provide a powerful deterrent to change for many farmers. Designed in an earlier era, they discourage crop diversity, crop rotations, certain soil conservation practices or reduced applications of pesticides. The chief offender is the federal commodity program, which covers nearly 70 percent of the nation's cropland. It awards subsidies according to the acreage a farmer devotes to a certain crop and yield. A corn farmer who wants to rotate some land to grow alfalfa and increase the fertility of the soil can expect a smaller subsidy.

The program also establishes base yields for farmers accord-ing to past production. Farmers strive for the highest base yields — however unrealistic they may be without heavy use of "off-farm inputs" — or lose potential income. Similarly, the system has encouraged farmers to expand onto marginally productive cropland to increase eligible base acreage. As a result, many farmers manage their farms to maximize subsidies, sometimes at the expense of environmental quality.

Other federal policies also discourage the use of alternatives. Government grading standards require fruits and vegetables to meet stringent cosmetic standards that have little, if any, bearing on nutritional quality. The "perfect" produce at our supermarkets comes at the cost of additional pesticides. Still others make it difficult to replace old pesticides with newer compounds that are safer and more effective. Overall, the current federal system of incentives is outdated.

Certainly, alternative farming is not a panacea. Even with a more favorable commodity support structure, farmers may find it difficult to meet increased management and labor needs and to devise new marketing strategies. More research is needed to identify the best techniques. Yet the potential benefits of alternative agriculture are too bountiful to continue to lie fallow. It is time we discarded outdated policies and practices and replaced them with new ones that are better for farmers, healthier for consumers, safer for the environment — and sustainable.

September 10, 1989

John Pesek *is head of the Agronomy Department at Iowa State University, Ames.*

★ ★ ★

The Paradox of Pesticides

Michael R. Taylor and Charles M. Benbrook

Pesticides are a paradox. They are essential to the agricultural economy, helping make possible the abundant, economical food supply Americans value. Yet they are also inherently toxic. Their purpose typically is to kill pests.

This dual nature — conferring benefits but posing risks — justifies the close regulation of pesticides by the Environmental Protection Agency (EPA) and state regulatory agencies. Many feel, however, that EPA's program is not working as well as it might in controlling residues of pesticides in food. In recent surveys consumers named pesticides as their No. 1 food-safety concern.

This concern is probably exaggerated — microbiological contamination of food is very likely a greater public-health threat — but, to its credit, EPA recently announced a plan to address an important part of the pesticide residue problem.

Much of the problem was inherited by EPA when it came into existence in 1970. Many of the pesticides of concern today were introduced in the 1950s and 1960s when testing requirements and safety standards were much less stringent. EPA is also saddled with pesticide laws that are inconsistent and difficult to administer.

In deciding whether a pesticide may be used or how much residue may remain on unprocessed food crops such as fresh tomatoes, EPA is directed by Congress to consider both the risks and benefits of the pesticide's use. When the food crop is processed, however — when the tomatoes become ketchup — EPA usually must drop this balancing approach. For pesticides found to cause cancer in laboratory animals, the law seems to adopt a zero-risk standard, precluding EPA from evaluating scientifically whether residues pose any real risk to human health.

At EPA's request, an expert committee of the National Research Council studied this problem and issued a report and recommendations last year. The committee found that EPA's inconsistent treatment of old and new pesticides was irrational from a scientific and public-health perspective. So, too, are the different standards it now applies to processed and unprocessed foods.

This situation works against the public health. EPA's application of a zero-risk standard for new pesticides in certain processed foods causes farmers to continue using older pesticides to which this standard has not been applied. Many of these older products are riskier than newer products whose danger is trivial or only theoretical. Yet it is the newer products that remain off the market.

To make the whole process more effective, our committee recommended that EPA apply a negligible-risk standard to old and new pesticides and residues in both processed and unprocessed foods. EPA has accepted such an approach in its new plan, a step that marks real progress toward making the pesticide regulatory system work.

Our analysis showed that the consistent application of a negligible-risk standard by EPA could reduce the potential risk of pesticide residues significantly. The usual benchmark for negligible risk is that a carcinogenic chemical must cause no more than one additional cancer death for every million people exposed to it. These risks are calculated by conservative methods that deliberately overestimate possible risks. In fact, the Food and Drug Administration has concluded that a negligible risk as described here is the functional equivalent of no risk at all.

Fully implementing this standard will also open the door to safer new pesticides and enable EPA to evaluate old ones more rationally. EPA will be able to retain in use those pesticides whose potential risks are insignificant and focus its efforts on reducing or eliminating exposure to pesticides whose potential risks are greater.

To take full advantage of this opportunity, EPA will need to increase the pace of its decision-making on both new and old pesticides and refine the role that benefits play in offsetting greater than negligible risks. EPA's greatest challenge will be dealing with its backlog of older pesticides. This requires a sustained commitment not only from EPA, but also from Congress, which will need to provide additional resources and perhaps new legislation.

Consumers, environmentalists, farmers and pesticide manufacturers all have a stake in the success of our nation's pesticide regulatory program. A successful program protects public health, promotes public confidence in the food supply and provides a more predictable business climate for those whose livelihood is agriculture. EPA's latest effort, while just a step along the way, is a positive move and deserves support.

November 1, 1988

Michael R. Taylor *practices law with King & Spalding in Washington, D.C., and was a member of the National Research Council committee that conducted the 1987 study on pesticide residues.* **Charles M. Benbrook** *is executive director of the Research Council's Board on Agriculture.*

★ ★ ★

Agriculture and Water Quality

Jan van Schilfgaarde

In 1982, ducks and other waterfowl began dying mysteriously at the Kesterson National Wildlife Refuge in central California. Biologists determined that water in the refuge was being contaminated by drainage from irrigated farms. The water contained substantial amounts of selenium, a usually benign trace element that can be deadly in high concentrations. The selenium had leached into the water from farm soil.

Dying birds are a distressing sight. Predictably, the situation caused an outcry in California and throughout the country. Farmers were forbidden to direct any more drainage water into the refuge and a cleanup effort was launched.

Sadly, however, that is not the end of the story for Kesterson, for California's rich San Joaquin Valley, for farmers in many parts of the West or for consumers nationwide. An expert committee of the National Research Council, which I chaired, found recently that similar situations are likely to occur elsewhere. Irrigated agriculture, which has yielded such bounty for the nation's dinner tables and wealth for U.S. farmers, is on a collision course with the environment in many areas.

It is a conflict more severe than most scientists had visualized and one politicians would like to avoid because of the wrenching social and economic choices involved. Yet, if further environmental catastrophes are to be avoided, a new water policy must be crafted that balances environmental, agricultural, and other costs and benefits more effectively than in the past.

Agriculture in the West has been dependent on irrigation for centuries. Today, nearly nine of every ten gallons of water in the region go to irrigate the fruit trees, vegetable fields and other crops destined for America's supermarkets. The price charged for this water often is far below the cost of its development. Current policies, established long ago, encourage farmers to waste water and grow crops on marginal land. Historically, farmers have not had to concern themselves with the environmental cost associated with the disposal of drainage water.

Irrigation agriculture depends for its survival on this disposal. It has long been known that drainage water often is saline, but the problem of disposal tended to be ignored. Now, however, we know drainage water can be deadly to wildlife and potentially to humans, as well. Although selenium was the main concern at Kesterson, other trace elements, such as boron and arsenic, also can be a problem. Like salts, these trace elements are dissolved by the irrigation water. As the water evaporates under the hot Western sun, they become concentrated and more toxic to plants and other organisms. Irrigation drainage also may contain residues of pesticides and fertilizers.

In the past, federal and local policymakers often have addressed drainage problems with temporary fixes rather than with permanent solutions, limiting themselves to options that allow agricultural production to continue unabated. Our committee felt it imperative that the full set of policy alternatives be placed on the table, however contentious they might be to some groups. These options should give appropriate weight to all interests, including environmental ones. Alternatives might include retiring land from agricultural production, raising water prices, and disposing of drainage water in the ocean where physically possible.

Certainly, current federal and state water quality regulations should be expanded to include irrigation drainage. There is no scientific reason why irrigated farmland should not be subject to water quality regulation, as is imposed on manufacturing plants and other types of industries.

Unfortunately, this is not a problem where "win-win" solutions exist to make everyone happy. Nor is it likely to be solved by a single government agency with a limited mission focus; a more multi-faceted approach is necessary. Possible solutions consist of a combination of technical options and changes in how society does business. None of these choices is cheap. All imply a tradeoff of values.

How to resolve these tradeoffs is a question more properly answered by elected officials rather than by a scientific committee like ours. Yet action is necessary if recurrences of Kesterson-like scenarios and potential adverse human-health effects from poor water quality are to be avoided. The national benefit from irrigated agriculture in the West is unquestionably great, but the laws of nature cannot be waived. When agriculture and the environment collide, something has to give.

November 26, 1989

Jan van Schilfgaarde *is associate director of the U.S. Agricultural Research Service in Fort Collins, Colo.*

★ ★ ★

Exploring the Mysteries of 'Deep Ecology'

James D. Nations

There's a movement called "deep ecology" among some environmentalists that could affect one of our planet's most precious and endangered resources: its biological diversity.

The movement is fascinating and noble — but it must be tempered with a more practical view of life on Earth. Here's why:

Plant and animal species in the world's tropical forests are being destroyed at incredible speed, in what some scientists have called an "extinction spasm." From the Amazon to Indonesia, species are being wiped out as their habitats are destroyed by farmers, loggers and others pushing into new territory. This terrible loss is eroding the genetic storehouse that scientists depend upon to develop new foods, medicines and industrial products. Chemicals and crops of the future are being lost forever.

Deep ecology deals with this problem by asserting that human beings have no right to bring other creatures to extinction or to play God by deciding which species serve us and should therefore be allowed to live. It rejects the anthropocentric view that humankind lies at the center of all that is worthwhile, saying instead that all living things — animals, plants, bacteria, viruses — have an equal inherent value.

As a conservationist, I am attracted to the core philosophy of deep ecology. Where I run into trouble with it, however, is in places like rural Central America or on the agricultural frontier in Ecuadorian Amazonia — places where human beings themselves are living on the edge of life. I have never tried to tell a Latin American farmer that he has no right to burn forest for farmland because the trees and wildlife are as inherently valuable as he and his children are. As an anthropologist and a father, I am not prepared to take on that job. You could call this the dilemma of deep ecology meeting the developing world.

When we in the industrial world stop to think about such a seemingly distant problem as this, we must resist the temptation to blame the poor farmers on the scene. The fact is that they generally understand the value of forest and wildlife better than we do in our society of microwave ovens and plastic money. They, not us, gather edible fruit, wild animals for protein, fiber for clothing and ropes, incense for religious ceremonies, natural insecticides, wood for houses, and medicinal plants.

I stayed once in southeastern Mexico with a Mayan farmer

who expressed his view this way: "The outsiders come into our forest and they cut the mahogany and kill the birds and burn everything. I think they hate the forest. But I plant my crops and weed them, and I watch the animals. . . . As for me, I guard the forest."

Today, that Mayan farmer lives in a small remnant of rainforest surrounded by the fields and cattle pastures of 100,000 immigrant colonists. The colonists are fine people who are quick to invite you to share their meager meal. But any of us in the industrial world who want to talk with them about protecting the biological diversity that still surrounds them had better be prepared to explain how it will affect them directly.

If you tell a frontier farmer that he must not clear forest or hunt in a wildlife reserve because this threatens the planet's biological diversity, he will politely perform the cultural equivalent of rolling his eyes and saying "Sure." The same can be said for the government planner in the nation where the pioneer farmer lives.

In other words, deep ecology makes interesting conversation over the seminar table, but it won't fly on the agricultural frontier of the Third World or in the boardrooms of international development banks.

Fortunately, even as advocates speak of deep ecology, many scientists and conservationists are showing a growing pragmatism about the situation. They are recognizing that to obtain the long-term benefits of biological diversity one must focus first — or at least simultaneously — on the immediate, short-term needs of individual people. Few wild gene pools are likely to survive intact in places where people have to struggle simply to provide their basic, daily needs.

This recognition must guide efforts to save the world's tropical forests. People on the frontiers of the developing world must receive material incentives that allow them to prosper by protecting biological diversity rather than by destroying it, and their welfare must be given equal consideration with the welfare of the forests. That done, we can return to the aesthetic arguments of deep ecology with the knowledge that, when we look up from our discussion, there will still be biological diversity left to experience and enjoy.

August 28, 1988

James D. Nations, *director of research at the Center for Human Ecology in Austin, Texas, presented a longer version of this article at a conference on biodiversity sponsored by the Smithsonian Institution and the National Academy of Sciences.*

★ ★ ★

4

THE NATION'S HEALTH

Food and Health

Arno G. Motulsky

Americans get bombarded with information about foods. Various food and food supplements taken to improve one's health are highly popular. Advisers to famous athletes tell us the virtues of certain dietary regimens.

Yet advice is often contradictory and the public has difficulty distinguishing scientifically based information from advice that has no solid basis.

What's a consumer to do?

Fortunately, we can answer that question with much greater confidence as a result of the widely publicized report on diet and health that was released earlier this month by the National Research Council. I chaired the committee of 19 experts that sifted through more than 5,000 scientific articles to prepare the report, which provides comprehensive information on the relationship between diet and chronic diseases such as heart disease, cancer and diabetes.

Although the report contained hundreds of statistics, the basic message to consumers was simple — namely, the current state of scientific knowledge allows certain conclusions about the relationship between diet and chronic illness. People *can* enhance their health by eating certain foods and avoiding others.

Conditions such as high blood pressure, coronary artery disease, obesity, osteoporosis and others are not caused just by diet or any other single factor; they are multifactorial in origin. A person's genetic makeup, for example, plays an important role in "predisposing" to all these conditions. Age and gender may further modify one's predisposition; for example, women are at lesser risk for conditions such as coronary heart disease but at higher risk for osteoporosis and iron deficiency.

Ideally, one would like to be able to determine the unique genetic-biochemical-metabolic makeup of every person to identify who is at highest risk of developing a given chronic disease. Based on this information, a "tailor-made" dietary prescription may be possible sometime in the future. In fact, we already can detect persons at high risk for coronary heart disease by measuring blood cholesterol.

For now, however, we lack the scientific knowledge necessary to identify those at high risk for most chronic diseases. And we know that the majority of chronic diseases occur in the general population — not only in those at high risk. Our committee therefore concluded that if the population as a whole adopted our dietary recommendations, a substantial reduction in chronic disease in the United States would result. The high-risk approach — when possible — and the population approach are complementary.

We found very strong evidence that saturated fat and cholesterol intake affect the frequency of arteriosclerotic coronary artery disease. There are also excellent data showing that salt intake increases blood pressure. The evidence for dietary factors causing various cancers was weaker but still quite suggestive.

We found no data to indicate that dietary supplements such as the vitamins-minerals taken by many people has any effect on preventing disease unless prescribed for specific medical indications.

The best diet, we concluded, is one that is moderate in total and saturated fat, high in complex carbohydrates and fiber, low in added sugars, and moderate in protein, especially animal protein. In practical terms, this means most Americans should eat more fish and less meat, and select leaner meats and trim off excess fat. They should select low-fat or nonfat milk and cheese, and eat more vegetables and fruits, especially green and yellow vegetables and citrus fruits.

Americans need to eat more cereals, breads and dried beans; use less oil and fat in cooking; and avoid fried food. They should restrict their consumption of egg yolks; salt; salty, smoked and preserved foods; and alcoholic beverages. Also, they ought to do a better job of checking labels for salt, fat and sugar content. We recommended these dietary patterns for children over the age of two as well as for adults.

To help Americans meet these goals, consumers should be provided more meaningful labels on their food purchases and restaurants should offer healthful menus. Along the same lines, nutritional education should be directed at all groups of society — not just the middle class.

The menu for a healthier lifestyle is clear. It's time more people put that knowledge into action.

March 21, 1989

Arno G. Motulsky *is professor of medicine and genetics at the University of Washington.*

★ ★ ★

The Spitting Image: Baseball Players and Chewing Tobacco

John C. Greene

A cheek full of chewing tobacco and a quick sideways spit. This image is as much a part of baseball as the peaked cap, the pinstriped uniform and the smack of bat against ball. The high visibility of smokeless tobacco use by the heroes in our national sport is cause for concern because baseball players are such prominent role models.

Many of their fans are picking up the wrong message. Within the general population, a shocking 16 percent of young men between the ages of 12 and 25 have used smokeless tobacco in the past year. One-third to one-half of these are regular users, and use is increasing.

During the last two years, I have helped lead a team of researchers from the School of Dentistry at the University of California at San Francisco that has conducted an extensive study of smokeless tobacco use and its health effects on baseball players. Our team gathered data to help convince players to give up tobacco for their own health — and for the sake of their fans. We studied and provided physical examinations of more than 1,100 players from major and minor league teams, including most of the players from the Giants, Athletics, Angels, Cubs, Indians, Mariners and Brewers.

Our results so far are very disturbing. More than half the players we examined in the first year were either current or past users of smokeless tobacco. Of the 40 percent who used snuff or chewed tobacco regularly, nearly half had lesions in their mouths. Snuff is taken as a pinch and placed inside the cheek or lips; chewing tobacco is used in a bigger wad.

All of the players with lesions were offered biopsies, and these all were benign. Yet these white or yellowish patches inside the mouth can become cancerous with long-term exposure to tobacco, usually after two or three decades of regular use. Users of smokeless tobacco products also have a higher incidence of periodontal disease and gum recession. In addition, as much nicotine enters the bloodstream from using smokeless tobacco as from chain-smoking cigarettes, which leads to other medical problems. One of our research team members calculated that a third of the nicotine from snuff and half the nicotine from chewing tobacco is swallowed, which may contribute to ulcers, digestive problems and cancer of the stomach.

In short, anyone who thought smokeless tobacco products were a safe substitute for cigarettes should think again.

Many players *are* concerned about the image they are projecting to young people. One key player cited his responsibility as a role model as his primary motivation for quitting the habit. When he spotted a 10-year-old youngster chewing and spitting while attending one of his community lectures, he asked why he was doing such an awful thing. "Because you do," the boy replied. The player then made a deal with the youngster, promising to quit if the boy would. The boy agreed, and the player hasn't used tobacco now for six years.

A pitcher for another team hasn't managed to kick the habit yet, but he now uses snuff only in private. In public, he chews bubble gum. Yet another player had help from another quarter. His wife, who had been nagging him to give up smokeless tobacco, greeted him at the front door one day with a lighted cigar in her mouth. That did the trick; he hasn't used tobacco products since.

Similarly, many baseball teams no longer allow manufacturers of smokeless tobacco to leave free samples in locker rooms. The San Francisco Giants, for example, provide free sunflower seeds instead.

In other words, the baseball community is beginning to come to grips with its smokeless tobacco problem. However, the dilemma for players still exists. Smokeless tobacco remains a powerful tradition and many players derive satisfaction from chewing while contemplating action. They also have difficulty breaking the addictive habit.

These professional baseball players need help in quitting smokeless tobacco use, and others should be discouraged from starting. It is simply unacceptable that professional players remain more than twice as likely as other young men to use smokeless tobacco. They should be showing their young fans how to hit and pitch, not how to chew and spit.

August 13, 1989

John C. Greene *is dean of the School of Dentistry at the University of California, San Francisco, and a member of the Institute of Medicine of the National Academy of Sciences.*

★ ★ ★

Clearing Our Vision About Alcohol Abuse

Robert D. Sparks

You are watching television when a commercial comes on for a local hospital's alcohol treatment center. Someone in your family needs treatment, the voice says. Pick up the phone and get them into our inpatient program. They need help — and so do you.

The commercial hits home. Your sister has a drinking problem and you want to help her. But, even as you write down the phone number, you begin to have doubts. Your sister is not a chronic alcoholic who staggers down the street. No, she usually goes to work and appears fine. Isn't a residential program a bit excessive? Who will care for her family while she's gone? And how will she pay for the treatment?

These concerns are familiar to many. The loved one for whom they feel concern may be a parent, spouse, child, sibling or friend. Or, they themselves may have the problem. Seventy percent of adult Americans drink alcohol, and many of them have at least occasional problems ranging from quarrelsome behavior to drunk driving. Yet, when people seek to solve these problems, they often feel overwhelmed by the prospect of enrolling in a residential treatment center. Their problem — too much wine with dinner, too many beers on the weekend, or binge drinking — may not seem severe enough to warrant such treatment.

I chaired an expert committee that recently examined this situation for the Institute of Medicine of the National Academy of Sciences. After studying medical reports and programs across the country, we concluded that no single approach works for everybody. Some people benefit from short-term counseling while others need specialized inpatient programs. Still others may overcome their problems without any formal treatment at all.

The challenge is to match individuals with the right programs. Yet, too often, the options available are not the ones needed.

Make no mistake; specialized hospital and residential treatment programs often do an outstanding job helping people to overcome alcoholism. As valuable as they are, however, they are not required for everyone. In fact, as many as a third of the people enrolled in hospital programs could be dealt with as effectively in programs that are much less costly. The same is true of Alcoholics Anonymous. It deserves all the praise it has received for helping alcoholics remain sober. Yet AA is not "the answer" for everyone who ever had a problem with alcohol.

Research has shown that many persons with less severe alcohol problems could change their behavior in several ways, such as through an intensive one-time counseling session or even by watching a video on drinking problems.

Does this mean society needs fewer hospital or residential treatment programs and AA meetings? No, it needs more of them. But it also must recognize that it is missing a tremendous opportunity by failing to broaden the base of treatment options for those now unserved.

Doing so would be a blessing not only for the individuals involved, but also for society generally. There is a widespread misconception that most car crashes, broken lives and other problems from alcohol are caused by the 10 percent of the adult population with serious drinking problems. However, the 60 percent of the adult population that consumes alcohol in more moderate amounts, by the sheer weight of its numbers, accounts for more problems. Offering them appropriate counseling or treatment options would reduce this toll.

It also would be a better way to allocate resources. Hospital programs typically cost several thousand dollars. When the treatment is appropriate and effective, this money is well spent. But for every two people treated successfully in this way, someone else might respond just as well, if not better, to a less expensive method. Insurance companies, government programs and others should be more flexible in paying for these options.

Overall, adopting a more flexible approach probably would be more costly because it would make treatment available to many more people. Yet, failing to do so is like saying that society will not treat people with broken fingers because it is too busy taking care of those with broken arms. The social cost of such a policy in terms of accidents, lost productivity and broken lives is simply unacceptable. People with alcohol problems need more treatment choices.

July 1, 1990

Robert D. Sparks *is president emeritus and senior consultant with the W.K. Kellogg Foundation.*

★ ★ ★

Needless Infertility

Howard W. Jones, Jr.

As the director of the nation's oldest center offering in vitro fertilization therapy for infertile couples, I weary of saying, "Sorry, your chances are not the best."

Our center has helped more than 500 couples bear one or more children. Yet an estimated 4.4 million women of childbearing age in the United States experience difficulty conceiving a child, and many of them might become pregnant with better treatment methods. Research to develop these methods is being hampered by a continuing moratorium on federal funding for human embryo and fetal research.

About 200 centers in the United States offer in vitro fertilization and embryo transfers to infertile couples. But success rates have been low, as was reported recently by a committee of the Institute of Medicine and the National Research Council on which I served. The procedure is appropriate as initial treatment for 10 percent to 15 percent of infertile couples, and in 1988 it resulted in a live birth only 12 percent of the time.

Drawing by Barbara Schreiber
The Atlanta Constitution, Ga.

Clinics like ours will never be able to help every infertile couple, but we could improve the odds considerably by learning more about the biology of human reproduction. Just as a brain surgeon studies brains, researchers in our field need to study sperm, eggs, pre-embryos, embryos and fetal tissue. Computer modeling, studies of animals and other methods can answer some of our questions, but we must investigate early prenatal development in humans directly if we are to truly understand it and learn why it does not always occur as desired.

Such research raises many difficult ethical and social issues, such as whether embryos are human, how "spare" embryos from fertilization procedures should be handled and whether fetal tissue from abortions ought to be used in research. These questions are profound, and they have been addressed in more than 85 reports from 25 countries.

Many nations have been more successful than ours in confronting the conflicting viewpoints and agreeing how to proceed. Americans have tended to polarize these issues rather than doing the hard work of devising research guidelines acceptable to most people. The Bush administration announced not long ago that it would continue to deny federal funding for human embryo and fetal research, and some members of Congress now are questioning the legality of the decision. Such wrangling needs to be replaced by people coming together in a less politicized manner, both within and outside government, to search for some consensus. As matters now stand, researchers are leaving the field and millions of infertile couples are frustrated.

From a scientific standpoint, the research opportunities have never been brighter. Our committee identified dozens of promising avenues of study, from the development of sperm and eggs to the process of embryo implantation. It listed several current clinical practices, such as stimulating and obtaining eggs from in vitro patients, that might be improved at clinical centers. It also called for closer investigation of technological advances such as freezing eggs and embryos.

A related scientific field illustrates what might be achieved with increased research on human reproduction. During the past two decades, scientists have made tremendous advances in techniques to assist conception in animals through in vitro fertilization and embryo transfer. Sharp ethical and scientific differences between humans and other animals prevent these methods from being applied directly to the treatment of human infertility. Yet the progress achieved with animals is pertinent, and it contrasts with the slow pace of research on human reproduction. Increasing this pace could lead not only to better treatment of infertility, but also to improvements in contraception, food production and efforts to sustain endangered species.

It is entirely proper in a democratic country that ethical questions involving science are resolved by society as a whole rather than by scientists alone. But, as one who confronts the heartbreak of infertility every day, I wish more Americans would recognize what we could accomplish by pursuing this research. Instead of saying "I'm sorry" so often, people like me could say, "Congratulations! You're pregnant."

June 3, 1990

Howard W. Jones, Jr., *is professor of obstetrics and gynecology at Eastern Virginia Medical School and director of the Jones Institute for Reproductive Medicine, in Norfolk.*

★ ★ ★

Who Is Going to Deliver Baby?

Roger J. Bulger

Having a baby should be one of life's most memorable experiences. For many American women, it's just that — but for the wrong reasons.

Some have a baby by Cesarean section but wonder later whether the procedure was really necessary. Others discover that their gynecologist has stopped delivering babies because of the fear of being sued and skyrocketing malpractice insurance rates. Still others lack health insurance and may have difficulty getting care at all.

Our country's obstetrical care system has delivered millions of healthy babies and saved countless young lives with its medical miracles. Yet it faces increasing problems, one of the most serious of which is the continuing controversy over medical professional liability.

Nowhere has the malpractice issue hit harder than in maternity care. Seventy percent of obstetricians-gynecologists reported in a 1987 survey that at least one claim had been filed against them. They were more than twice as likely as other physicians to be sued; their annual malpractice insurance premiums in some cities now exceed $100,000.

A committee of the Institute of Medicine of the National Academy of Sciences concluded in a study released this past week that the malpractice situation helps explain why so many women across the country, especially those from rural areas and inner cities, cannot find anyone to deliver their babies.

Obstetricians generally practice in metropolitan areas; in rural areas, most babies are delivered by family physicians. About 2,500 certified nurse-midwives also practice across the country. Significant numbers in each of these groups have eliminated or limited their obstetrical practice because of liability concerns.

The trend has made it difficult even for some women with adequate health insurance to obtain obstetric care. Low-income women, however, have been hurt the worst. Not only do they generally lack health insurance, but they also are more likely to smoke, have poor diets and otherwise be at risk of experiencing complications. Many obstetrical providers have cut back on services to these high-risk women and reduced their Medicaid caseloads because they fear being sued for any bad outcomes, even those that are unavoidable.

The day-to-day practice of obstetrics also has changed. Some of these changes have been beneficial, but others are more questionable. Notably, there is considerable evidence that routine use of electronic fetal monitoring in normal pregnancies is both expensive and unnecessary, occasionally leading to inappropriate Cesarean sections. When a defective baby *is* born, a physician is far less likely to be sued for performing an unnecessary C-section than for not doing one. Obstetric technologies, such as fetal monitors, should be assessed more systematically before they become accepted as standard practice.

Whether high malpractice insurance premiums constitute a real economic burden for obstetrical providers is open to debate. But what is certain is that the premiums will remain high so long as obstetrical disputes are settled through our existing legal structure. The tort system now in place has resulted in increased costs, reduced availability of services and an erosion in trust between doctor and patient.

Several states have begun experimenting with alternatives to tort law, such as by providing no-fault compensation for certain birth defects. These experiments offer one way of easing the malpractice dilemma in obstetrics; similar efforts are needed around the country.

The most immediate need of all, however, is to provide uninsured women with greater access to maternity services. Extending additional immunity to physicians at government-financed clinics, contributing to liability coverage for Medicaid providers, and expanding the National Health Service Corps are possible ways to encourage doctors to provide obstetrical services.

The current situation in obstetrics cannot be allowed to persist until the malpractice issue is resolved for medicine generally, a process that could take many years. Too many women and babies are not being served. States need to address immediately the deterioration in maternity services for the poor that have been worsened by liability concerns.

Having a baby should be memorable for the right reasons. It's time we gave birth to a maternity-care system that makes this possible for many more American women.

October 15, 1989

Roger J. Bulger, *president of the Association of Academic Health Centers, chaired a committee of the Institute of Medicine of the National Academy of Sciences that studied medical professional liability and the delivery of obstetrical care.*

★ ★ ★

Accidents Are Not Always Accidental

Susan S. Gallagher

They are the deaths that fill our local evening news: A family suffocated in a fire. A child drowned in a neighbor's swimming pool. The teenager who shot himself. The young mother crushed in a car accident.

Events like these are so common they appear inevitable. But suppose the mother's car had been equipped with an airbag? What if the teenager's parents had locked up their gun, the pool had been surrounded by a locked fence, and the family had installed a smoke detector?

Every year about 150,000 people in the United States die of injuries. One in four Americans is injured annually. This is a staggering toll, yet accidents often are perceived — incorrectly — as random events that cannot be anticipated or prevented. Case in point: After the recent series of crashes, deaths and injuries on military ships, headlines read "Navy Calls Accidents String of Bad Luck."

Injuries are so common that they lack the aura of scientific mystique associated with cancer or AIDS. When we hear about someone receiving an organ transplant, we are fascinated by the patient's story. Rarely do we ask whether the death of the organ donor was due to injury and might have been prevented.

Similarly, when people are severely injured, their families usually focus on the need for rehabilitation and long-term care — not on preventing similar accidents. Contrast that with the tradition of supporting medical research on heart disease, cancer or whatever disease killed a loved one. Most important, a life lost to injury is almost always obvious and mourned. But it is rarely evident — and celebrated — when a life is saved through prevention efforts.

Dramatic progress has been achieved during the past three decades in our ability to prevent injuries and mitigate their impact. Experts in epidemiology, medicine, engineering, biomechanics and other disciplines have shown how to save tens of thousands of lives each year. But neither scientists nor the larger community have mustered the political will to apply this knowledge and implement proven strategies.

Congress, responding to a report of the National Research Council and Institute of Medicine on this needless suffering, established a unit at the Centers for Disease Control to oversee research, practice and training on injury control. Last year it more than doubled the program's budget to $21 million. These steps deserve applause but must be expanded drastically to tackle a problem costing society $158 billion a year.

A decade ago 26 states rescinded laws requiring motorcyclists to wear helmets; motorcycle fatalities increased 40 percent. Airbag technology has been available for 20 years; only now are manufacturers introducing airbags in automobiles. Technology also exists to make self-extinguishing cigarettes, which could prevent an estimated 2,000 deaths and 6,000 burn injuries annually. But legislation requiring manufacturers to produce them has stalled. Even simple and inexpensive measures such as smoke detectors have not been universally implemented.

Some existing safety efforts, meanwhile, have not been proven effective. Research suggests that traditional driver education programs have not been as useful as portrayed; some alternatives, such as graduated licensure of beginning drivers, should be tried. Better programs also are needed to reduce cases of lead poisoning in children and to make amateur competitive sports safer.

Injuries never will be eliminated. But consumer products and physical environments could be made less hazardous, such as with designated traffic lanes for bicyclists or with detachable bases to reduce baseball sliding injuries. Stronger safety standards in automobiles could be enacted and existing ones enforced more vigorously. More could be done to change social norms involving dangerous behaviors, as has been done with drunk driving. Litigation can help pressure manufacturers to stop making unsafe products. Most needed are new injury control programs on the local level, training programs for professionals, and better data systems to identify where the problems are.

In other words, we need to stop accepting injuries as a fact of life — and death. As former Surgeon General C. Everett Koop pointed out, "If some infectious disease came along that affected one out of every four children in the United States, there would be a huge outcry and we would be told to spare no expense to find the cure — and to be quick about it."

August 19, 1990

Susan S. Gallagher *is a senior scientist at Education Development Center Inc. in Newton, Mass. This article is adapted from a longer version she co-authored with Ilana Lescohier and Bernard Guyer in* Issues in Science and Technology.

★ ★ ★

Changing Behavior to Limit the Spread of AIDS

Heather Miller and Marshall Becker

Anyone who has tried to stop smoking knows it is not easy to change a habit that is pleasurable but dangerous. Yet some people *do* give up smoking and other behaviors involving fundamental urges, compulsions and addictions. Their success deserves more attention as the AIDS epidemic continues to claim lives.

Theoretically, the spread of the HIV virus that causes AIDS could be virtually stopped by altering the behaviors known to transmit it: unsafe sex and intravenous drug use. Over the past several decades, social and behavioral scientists have accumulated valuable knowledge about modifying unhealthy behaviors. Their approaches and insights ought to be applied to altering AIDS-related behaviors, as appropriate.

All too often, however, drug use and unsafe sex are portrayed in moralistic terms. People are told to "just say no," even after it's clear they are unwilling or unable to do so. Clinging to such a simplistic approach in the face of deadly consequences and the availability of more effective strategies exhibits moral mean-spiritedness.

One lesson from the behavioral and social sciences that might be applied to the AIDS epidemic is that many people respond more effectively to a choice of options than to a single, dictated course of action. The evidence also shows that most individuals are better at modifying selected behaviors than at changing their entire lifestyle.

For example, many dieters find it easier to reduce the size of portions than to give up certain foods altogether. Similarly, some intravenous drug users who cannot obtain or accept treatment immediately, for whatever reason, may find it easier to stop sharing drug paraphernalia or to sterilize injection equipment than to practice abstinence.

To be effective, behavioral intervention programs need to consider people's health beliefs and perceptions of their own efficacy. The programs should incorporate knowledge about how people make decisions and why they take risks. Efforts to promote behavioral change also should be based on current theories of relapse prevention, social norms, community organization and environmental restructuring.

Studies have shown the importance of providing accurate information in language that the intended audience can understand, delivered through the mass media and other routes that take reading skills into account. Messages should be realistic, based neither on moral judgments nor on excessive levels of fear. In the case of AIDS, this means overcoming any reluctance to discuss sexual behavior in a clear, explicit and comprehensible fashion.

Because our society is diverse, interventions that are effective for one group may not work with others. Therefore, multiple strategies are needed to help different groups, and they must be repeated often — both to remind people about the risk of AIDS and to reinforce desired behavioral changes.

Regardless of what we do in the future, many people already are involved in high-risk behaviors, and some sexual and drug experimentation is likely to continue. Data indicate that the majority of Americans become sexually active in their teens. Rather than engaging in wishful thinking, we must find the courage to heed these data and begin the difficult process of discussing sex and drug use with adolescents in a manner that will help them act to protect themselves. Any program that fails to go beyond exhortation and come to grips with this need consigns millions of Americans to possible death.

Similarly, we must be prepared to explore innovative strategies, such as needle-exchange programs, to save the lives of those who are not yet ready to elect — or who cannot obtain — more traditional treatment for their drug problem.

We are eight years into the epidemic and there is no indication that an AIDS vaccine or cure is near. Given the dire consequences of this disease, public officials, health professionals and others must begin making better use of findings from the behavioral sciences. They also must make a substantial commitment to research efforts that will expand this understanding rapidly.

We have identified specific behaviors that can transmit AIDS. It's time we put to use our knowledge of what works best in facilitating and sustaining changes in these risky behaviors.

November 12, 1989

Heather Miller *is a study director of the National Research Council's Committee on AIDS Research and the Behavioral, Social and Statistical Sciences.* **Marshall Becker**, *professor of health behavior and associate dean of the School of Public Health at the University of Michigan, serves on the committee.*

★ ★ ★

The Dilemma of AIDS Drug Experiments

Robin Weiss and Theodore Cooper

One of the most difficult aspects of the AIDS epidemic for medical researchers is having to turn down requests from AIDS patients who volunteer to participate in experiments of new drugs.

From the researcher's perspective, the applicant may fail to meet certain criteria, such as a prescribed set of symptoms, and therefore be ineligible. Yet turning down such requests from dying patients may appear cold and insensitive. After all, if the patients are likely to die, why not let them at least *try* the drug? It could help them and, if nothing else, it will give them hope. Even if the drug fails, its use may produce knowledge to help others. What, then, is the harm in letting people participate?

As we learned during our recent study of AIDS and HIV infection for the Institute of Medicine and the National Academy of Sciences, medical researchers, physicians, AIDS patients and others have been struggling with the ethics of this difficult situation. Our committee concluded that the most compassionate course is for drug trials to continue to follow rigorous scientific guidelines.

Experiments to compare one treatment with another in humans are called clinical trials. Intuitively, it seems a simple matter to administer a drug to a patient and then see if the patient improves. However, unless the drug has dramatic results, it can be difficult to tell how it has affected the person's condition when the illness itself has spontaneous ups and downs. If two different drugs are given to two people with the same disease, differences in outcome may be due more to differences between the two people than between the two drugs.

Over the years, scientists have sought to diminish these kinds of ambiguities by developing strict guidelines for drug experiments. When comparing two or more drugs, for example, scientists take care to follow strict eligibility criteria when they recruit volunteers, assigning patients randomly to different groups. The groups are made as similar as possible, but they receive different treatments. In many drug trials, the control group receives a placebo. For most AIDS drug trials involving persons with symptoms, however, the control group should receive the drug AZT, which has proven effective. This way everyone has a chance to benefit from therapy.

Any difference between the groups may affect the validity of the results. For instance, if the patients in one group happen to have less severe illness at the beginning of the experiment, then this, rather than the experimental drug, may account for an improved outcome at the end of the test.

Another rule of clinical trials is to try to use the right number of test subjects. This is done by calculating the smallest number of people needed to show a statistically significant difference between the experimental groups. The goal is to avoid exposing more people than necessary to a potentially toxic drug and to complete the study in the shortest possible time.

This leads to one of the most crucial points about clinical trials, which is that they are *experiments*, not treatment. Three things can happen when testing a new drug: it can make no difference in the patient's health; it can be beneficial; or it can make the patient worse by accelerating disease or exerting toxic effects that make the patient sicker. In the case of AIDS drugs, it is frequently overlooked that people can be made sicker by experimental therapy. This is what happened with the drug suramin, which looked promising but proved in clinical trials to have serious toxicity and no clear benefit.

Allowing wide distribution of unproven therapies or relaxing entry criteria for AIDS drug trials, therefore, is not a benign action. Trials that are larger than necessary or sloppy in design and execution will delay the effort to find meaningful therapy for HIV infection and AIDS.

A number of researchers have been studying alternatives to traditional randomized trials, and their work warrants close monitoring. The suffering caused by AIDS has been so severe that innovations must be pursued vigorously and any effective new drugs made available as widely and quickly as possible, perhaps through treatment demonstration projects. The federal government needs to provide more funding to support clinical trials, and it should work with industry to develop better systems for gathering, analyzing and sharing test results.

Our committee also called on researchers to make a greater effort to extend their trials to a wider variety of new drugs and to such untapped populations as women, IV drug abusers and pediatric patients.

In the end, however, the first rule of these experiments must continue to be that they are sound scientifically. Doing so is actually the most humane — if often difficult — course of action if we are to bring this terrible epidemic under control as soon as possible.

October 4, 1988

Robin Weiss, *director of AIDS activities at the Institute of Medicine, was study director of a committee of the Institute and the National Academy of Sciences that examined the AIDS crisis.* **Theodore Cooper**, *chairman and chief executive officer of The Upjohn Company, chaired the committee.*

★ ★ ★

Identifying What Works in Medicine

Samuel O. Thier

The U.S. national health care bill exceeds $600 billion annually. But many are questioning whether the care — including diagnostic tests, surgical procedures, drug therapies and other medical interventions — always is truly effective.

Consider hip fractures, which an Institute of Medicine committee recently examined. About 260,000 Americans, mostly elderly, suffer hip fractures each year. As many as half of them require care for the rest of their lives, at great expense to their families and society. Direct medical care costs total about $6 billion per year. Yet no clear consensus exists among physicians about certain aspects of the treatment of hip fractures, such as how long patients should be hospitalized, which surgical options to use, or the most effective sequence of surgical interventions.

Vast anecdotal evidence exists about dealing with hip fractures, but the plural of anecdotes is not data. A more rigorous evaluation might analyze how patients have responded to different kinds of treatment, the number of days they were hospitalized, mortality rates, costs and a host of other questions. Focusing payment on the most effective approaches might reduce costs. Cutting the bill by even a small fraction could yield substantial savings, perhaps freeing up resources for other needs such as prenatal care, drug treatment, basic care for the uninsured or AIDS. Even if no money were saved, care would be improved.

So, too, across the spectrum of medical care. Studies have called into question whether coronary bypass operations, Cesarean sections and other common procedures are overused. Some procedures that are useful in specific situations have become accepted across the board — for example, the use of electronic fetal monitoring in obstetrics. Although the value of this technology in normal pregnancies is unproven — and perhaps even negative — clinicians feel pressure to apply it generally. They risk being accused of negligence if they fail to use the devices and something goes wrong.

Too often, care is unnecessary or inappropriate, or appropriate care and preventive services are not provided. Accepted medical practices vary from city to city. Even within a hospital, one physician may be much more likely than another to order a costly test or procedure even though evidence of its effectiveness is lacking. Unnecessary expenditures are rarely the result of deliberate fraud, and professional autonomy is important for physicians. But as the cost and complexity of choice in any clinical circumstance continue to expand, physicians need to know what works.

This presents both a challenge and an opportunity for the medical community and society generally. Our institute has been studying medical practices for treating hip fractures, breast cancer and heart disorders, and others are looking at chronic pain, bedsores, cataracts, depression, sickle-cell disease, prostate enlargement and other conditions. Yet even with these studies and increased interest among federal officials, the knowledge base is likely to remain meager for years to come. And it will not be easy to translate findings into actual changes in the day-to-day practice of medicine.

Evaluating medical procedures more systematically is one of the best — and most overlooked — ways of improving health care. In recent years, our country's familiar health care system has been swept by a host of changes, from proliferating health maintenance organizations to hospital emergency rooms overwhelmed by uninsured patients. Some valuable changes have improved quality and efficiency while controlling costs. Yet many efforts have focused on either cutting benefits or limiting patient access to the system, raising troubling questions of equity and quality.

Studying the effectiveness of our actions offers a promising alternative for controlling costs while, at the same time, improving care. Medical professionals should participate actively in this process, viewing it as a way to improve their craft rather than as an intrusion on their prerogatives. Medicine is a learned profession, and one of the responsibilities ceded by society to any learned profession is to set and enforce standards. Although the diversity of human beings makes it impossible to create a single rule book for clinical practice, more can be done to separate the accepted from the truly effective.

If we physicians do not take the lead in doing this, then others with less training will do it for us. As health care costs approach $2 billion a day, Americans need to know that everything being done in their hospitals and doctors' offices is truly worthwhile.

October 28, 1990

Samuel O. Thier *is president of the Institute of Medicine of the National Academy of Sciences.*

★ ★ ★

5

MAKING SENSE OF SOCIAL PROBLEMS

Confronting the Facts in American Race Relations

Gerald David Jaynes and Robin M. Williams, Jr.

Many white Americans believe the civil rights era of the 1960s removed all barriers to equal opportunity and that blacks are making steady gains towards equality.

The facts suggest otherwise. We recently led a team of scholars that carried out the most comprehensive review since World War II of the status of black Americans. After reviewing a wealth of data and research, we were persuaded that time alone is unlikely to resolve our nation's racial problems.

The data show that, after several decades of rising economic status relative to whites, blacks' gains have stagnated or declined on many measures since the early 1970s. Blacks remain substantially behind whites in health and life expectancy, educational and residential opportunities, and political participation. Throughout the 1980s, about two-fifths of all black children lived in poverty during any given year.

These findings have important implications for public policy. The large gaps between blacks and whites — which have remained constant or widened during recent years — discredit the view that racial equality can be achieved without planned government actions. To have no policy is, in fact, to have a negative policy in the area of race relations.

Reasonable people may agree that a policy is needed without agreeing on the policy itself. Much debate has focused on whether policies are needed most to change black people's

Drawing by Pat Bagley
The Salt Lake Tribune, Utah

behavior or, instead, to expand their opportunities. Those who would change behavior often hypothesize that a special "culture of poverty" exists among poor blacks and must be altered if people are to begin living more productively. However, we found little evidence to support this hypothesis.

Does this mean that poor blacks do not engage in behaviors that help perpetuate their poverty? No — crime, teenage pregnancy and the like can be found in any inner city. However, there is little in the record to suggest that these behaviors occur independently of socioeconomic conditions — which can be changed with social policies.

Black-white cultural differences have narrowed, not widened, since 1960. Blacks respond to changes in American society much as whites do. Their poverty rates generally rise and fall with those of whites. Black employment rates, educational attainment and births to unwed mothers do likewise.

Furthermore, we found that much racial inequality continues to be due to discrimination and the segregation of poor blacks from quality schools, neighborhoods and other social institutions.

These findings lead to the clear conclusion that the main thing that needs enhancement is not black attitudes but black opportunities. Cultural values and beliefs are certainly important, but the fundamental problem of poor, black Americans — as of all poor Americans — is their lack of opportunity and their underlying poverty. A large body of research literature shows, for example, that employment programs that improve poor people's skills have been more successful than those that seek to inculcate "discipline" among the poor by placing them in low-paying jobs devoid of meaningful training.

The record is also clear that previous programs by governments and private institutions have made a large difference in the opportunities and conditions of black Americans. These programs have included the Job Corps, Head Start, financial aid for college students and health services for young mothers.

Social policies with demonstrated benefits include the provision of education, health care and other services to enhance people's skills and productive capabilities. In addition, discrimination and involuntary segregation must be reduced. Better income maintenance and other family-assistance social welfare programs are needed to avoid long-term poverty. Above all, what black Americans need is full employment.

Certainly, anyone except an unrepentant segregationist can take pride in the gains our country has made since the days when blacks were forced to ride in the back of the bus. Over the past 50 years, the social status of black Americans has on average improved dramatically. Yet the many problems that remain make it clear that our national agenda in race relations remains unfinished.

Further progress requires public and private programs to enhance the productive capabilities and the opportunities of the poor. Our nation need not debate endlessly whether such efforts can make a positive difference. The empirical record shows that they can, and now is the time to put this knowledge into action.

September 17, 1989

Gerald David Jaynes *of Yale University and* ***Robin M. Williams, Jr.****, of Cornell University are the editors of the recent National Research Council report* A Common Destiny: Blacks and American Society.

★ ★ ★

Homeless Children: An Emerging Tragedy

Ellen L. Bassuk

At least 100,000 American children will go to sleep tonight — or try to sleep — on sidewalks, in shelters or wherever they can. Each of these homeless children has a uniquely tragic story to tell, but the stories add up to a chaotic tale of Dickensian proportions.

I served recently on a committee of the Institute of Medicine of the National Academy of Sciences that found the growing number of homeless children to be "nothing short of a national disgrace."

Our report echoed what has become a growing body of research on homeless children that confirms some of our worst fears about their future. In a recent study I helped conduct in Massachusetts, a majority of school-age children from homeless families in shelters agreed with the statement: "I think about killing myself, but I would not."

Many of the preschoolers exhibited severe and multiple developmental lags. Compared with poor children who did have homes, these youngsters had serious problems with language development, gross motor skills, and social and personal development. For example, one 19-month-old boy we saw stopped eating and developed nightmares when his family moved into a shelter. Despite the obvious needs of such children, however, many of them receive few services.

As winter settles over the country the plight of homeless children becomes temporarily more visible, even though their year-round struggle is always desperate. Families are the fastest-growing segment of the homeless population, now making up about one-third of the total. Most of these families are headed by a woman with two or three children, a majority of whom are 5 years old or younger.

Emergency shelters and substandard welfare hotels are poor substitutes for the safety and security of a stable home. Small, crowded rooms with little privacy and, in many places, no dining facilities are inadequate places to raise children. This is particularly true because so many mothers are understandably focused on daily survival and have little energy left to respond empathetically to a distressed child.

Many homeless children rarely go to school or attend only erratically. Transportation is frequently inaccessible or schools are located great distances from shelters. One homeless mother described to me how she had to ride public transportation for more than two hours every morning to take her two older children to schools located in different neighborhoods far from their shelter.

Her case was extreme, but a more common example is illustrated by a homeless mother of three children who had been living in an emergency shelter for six months. Over the past five years her family had moved eight times, doubling up in overcrowded apartments and various welfare hotels. Her three children attended school irregularly and all had repeated a grade. The oldest was failing two subjects and the middle one had memory and speech problems.

Stories like these are not isolated examples. The lives of homeless children are filled with fear, chaos, insecurity and loss. Because their daily activities are not routine or reassuring they do not have the emotional anchors many children take for granted.

What is so discouraging about the situation is that we know ways to ease or solve it for many children. The greatest needs, of course, are for decent, affordable housing and supports to help poor families maintain their homes. Also, various programs should be expanded so that they can be made available to homeless preschoolers. Infant stimulation, enriched day care and Head Start programs are among those that have been shown to make a measurable difference in the lives of severely disadvantaged young children.

Research has also shown that many of the developmental delays brought on by homelessness are reversible. Some of the damage can be undone by providing children with the homes and support they need. However, ignoring the problem today will only result in a higher price tomorrow. In other words, we *can* act as a society to help homeless children. Successful programs exist that can be expanded to reach more of these youngsters who are at risk of becoming seriously anxious and depressed, failing at school and losing hope. That has happened to too many homeless children already. We must resolve that this new year will be different.

January 3, 1989

Ellen L. Bassuk *is associate professor of psychiatry at the Harvard Medical School and president of the Better Homes Foundation.*

* * *

The Gender Wage Gap

Robert T. Michael

Women in the United States earn only about 65–70 cents for every $1 earned by men. Is this due to bias against women — or to differences between women and men in their levels of education, work experience and other measurable factors?

That is the central question behind the continuing — and continually controversial — debate in our country about how to achieve "pay equity" and over whether "comparable worth" is a worthwhile strategy. Comparable worth seeks to eliminate gender bias from the labor market by employing techniques of job evaluation that are designed to use objective criteria to value the content and requirements of jobs, whether those of nurses, truck drivers, teachers or plumbers. The goal is to establish wages that reflect a job's true worth rather than historical salary patterns.

It is a debate subject to hyperbole on both sides, but the outcome is important because the concept could affect the paychecks of millions of Americans. The primary argument from proponents of comparable worth has been that it provides fairer pay for jobs held mostly by women. The main opposing argument has been that it imposes an arbitrary assessment of a job's relative value, disrupting the workings of the labor market and possibly causing inefficiency and inequity.

One reason the debate has been so contentious has been a lack of conclusive evidence about the heart of the matter, which is whether pay differences *really are* due to systematic bias against "women's jobs." Comparable worth is designed to address systematic undervaluation of jobs held by women. If, however, the problem is that women are prevented from entering male-dominated jobs, then the proper remedy may be equal-opportunity legislation. Of course, both problems can occur simultaneously.

A recent series of studies on pay equity from the National Research Council helped clarify the situation. It showed that the wage gap is indisputable. The studies confirmed that women receive lower pay than men across a wide range of occupations and employers. A discrepancy remains even after accounting for education, job experience and other measurable differences between men and women.

Whether this wage gap is attributable to systematic bias against women or to other market forces still poorly understood or measured, however, remains unclear. A number of studies have found that wage differences by gender are diminishing, although gradually. It also appears that although jobs held mostly by women do receive lower wages than those held mostly by men, women generally are not paid less for the same job.

One study found that starting salaries for specific civil service positions fell when women or minorities entered those jobs in large numbers, even though the job requirements remained the same. This suggests the jobs were devalued *because* women and minorities held them, as comparable-worth advocates have argued. Yet, even here, alternative explanations are possible.

Prescribing the medicine of comparable worth in the face of such an inexact diagnosis is a political question rather than a research subject. Yet the recent experience of several states and localities suggests that doing so is less traumatic than some have predicted.

A comparable-worth plan introduced in Iowa's state employment system in 1985 cut that state's gender wage gap in half, from 6 percent to 3 percent. Minnesota appears to have had success with similar legislation. In Australia, the relative wages of women were raised substantially with no evidence of adverse employment effects or disruptive economic effects.

A number of studies suggest that pay equity plans designed for specific employers in the United States can, at most, correct about 20 percent of the observed gender wage gap. They also have little power to address differences across firms or industries, and it is unclear whether they will affect worker productivity or overall unemployment. Yet they do provide a substantial increase in salaries for women workers and appear not to cause negative side effects, at least in the short run.

It is essential that society debate comparable worth on the basis of all available facts and that it find empirical answers for the remaining questions. One of the fortunate aspects of the controversy is that it lends itself to objective analysis. A just and efficient wage structure for working Americans can — and should — be based on an understanding of the complex reality of what happens in the workplace and in the job market.

October 22, 1989

Robert T. Michael, *director of the National Opinion Research Center and professor of education and public policy at the University of Chicago, chaired a National Research Council panel on pay equity research.*

★ ★ ★

Child Care in Disarray

John L. Palmer

It is late afternoon and two women are picking up their 4-year-old children. The first woman, a married attorney, drives to an attractive child care center to get her child. The second woman, a single mother who works as a waitress, takes the bus to a friend's one-bedroom apartment, where her child spends the day with five other neighborhood children.

If you were in a position to assist one of these two women, whom would you choose?

Uncle Sam now favors the attorney. She claims a tax credit on her income tax and has a Flexible Spending Account with her firm that lets her pay the child care center with tax-free dollars. The waitress, who earns the minimum wage, is too poor to use either benefit. Like four of every five low-income children, her child cannot get into a Head Start program.

Of course, as any working parent knows, the balance sheet on child care is not defined by money alone. There also is the emotional cost of sending a child to school with a fever or of missing the child's class play because of a business meeting.

But, as a panel I chaired for the National Research Council reported recently, the federal government has been providing less and less support to poor, working parents, who need help most. In 1972, 80 percent of federal child care support was targeted to low-income families through subsidies to child care centers and other providers. By last year, these programs accounted for less than 30 percent of the total as federal support had shifted to direct consumer subsidies primarily benefiting middle- and upper-income families.

Our panel did not take a position on the child care bills now before Congress. But we did urge that special attention be paid to the millions of poor, working parents who lack the resources to place their children in safe, healthy and enriching settings.

Children from poor families are especially at risk, but they are hardly alone. Most children in the United States now have working mothers; all too often, quality child care is unavailable. Many families find the care available from neighbors or local centers to be too costly — or simply inadequate. Parents who work non-traditional hours, or those whose children have medical problems, often have special difficulty arranging child care services.

Not that white-collar professionals have it easy. They may worry less about paying the bills, but they also find child care to be a source of stress. Organized programs for infants and toddlers are scarce in many neighborhoods, and an estimated 2.1 million youngsters are "latchkey kids."

A growing body of research confirms that quality care is essential to a young child's healthy development, but our national child care system remains fragmented and inadequate. Services have developed without any rational framework of legislation or policy. State regulations vary widely and often ignore recognized standards of quality. Some states allow a single person to care for as many as 12 infants.

Legislative strategies differ, but several steps clearly are required to improve the quality, affordability and accessibility of child care. The need to assist poorer families and to expand compensatory programs such as Head Start is acute. National standards for the quality of all child care should be set. In addition, most parents should have their jobs protected for up to one year if they choose to stay home and care for a newborn. Research indicates it is important that new families get off to a good start. Other countries have successfully implemented parental leave policies.

It would be foolish to legislate a single kind of child care for all children. One good thing about the current system is its diversity. But parents should be able to choose among options of high quality, including providing care themselves, rather than lurching from crisis to crisis.

The price tag for quality care will not be low. For example, our panel recommended that government at all levels provide an additional $5 billion to $10 billion annually just as a first step. But attorneys, waitresses and millions of other parents can attest that our children deserve better, and the future of our country depends upon it.

April 1, 1990

John L. Palmer *is dean of the Maxwell School of Citizenship and Public Affairs at Syracuse University.*

* * *

Effective Drug Treatment

Lawrence S. Lewin and Dean R. Gerstein

Despite the deficit, President Bush and Congress are proposing large increases in funding for drug treatment programs. Why? Because publicly supported drug treatment is the neglected front in the national drug war.

More than 2 percent of the U.S. population over age 12, or 5.5 million people, clearly or probably need drug treatment. Fewer than one in six now get it. There are 66,000 people on waiting lists for openings at public drug treatment programs across the country; many more don't apply. An estimated 105,000 pregnant women have serious drug problems, as do a million people under the supervision of the criminal justice system.

These numbers are growing. Although surveys indicate that casual drug use is decreasing, severe drug abuse and dependence are becoming more prevalent, the menu of illegal drugs is expanding, and the AIDS epidemic makes treatment even more complex. Yet, until recently, drug treatment was regarded too lightly by most officials. Public funds were aimed at drug busts and other law enforcement efforts. Only about one-seventh of the $3.9 billion allotted in 1989 for the federal anti-drug campaign went for drug treatment.

But would increased spending on drug treatment really lead significantly more people to end or even reduce their dependence on illegal drugs? If done right, the answer is yes.

We led a committee of the Institute of Medicine of the National Academy of Sciences that recently studied drug treatment programs, and we found that the better ones not only can get addicts off the street but also help change their lives. The benefits to society in reduced crime, improved productivity and the like exceed the costs. Expanding treatment is both humane and cost effective.

Not every drug user needs formal treatment. Many experimental or casual users just lose interest in drugs or respond to preventive counseling or disciplinary sanctions. Drug treatment is justified and appropriate when an individual shows clinically significant signs of dependence or chronic abuse.

Not all treatment programs are the same; quite the contrary. Our country's treatment delivery system is sharply divided between public and private tiers. Individuals with private insurance generally enter hospital-based programs for inpatient detoxification and treatment. Those whose treatment is paid from public funds are more likely to make weekly visits to a clinic or to enter a spartan residential community, with a daily program of work duties and intensive group therapy. Public programs that maintain addicts on methadone also are common.

Surprisingly, the costs of these different approaches are not based on their effectiveness, although there is a direct relationship between the length of time spent in treatment and long-term effectiveness. Although objective data are sparse, largely because research and evaluation programs were slashed several years ago, there is good evidence that methadone maintenance for heroin dependence and residential "therapeutic communities" are cost effective. The evidence is scantier for outpatient non-methadone treatment, but the much lower costs make even modest benefits worthwhile.

Other treatment approaches, such as self-help groups and prison-based "boot camps," have not been evaluated in sufficient detail. Evidence of effectiveness also is generally slim for the programs offered by private providers. There is no evidence that expensive hospital-based chemical dependency programs are more effective than similar programs not sited in hospitals. Private payers should insist on reliable evidence of clinical outcomes and not waste money on unnecessary services or overhead charges.

One big concern about drug treatment programs has been whether they have any impact on patients who enter treatment in the community to avoid prison. An encouraging finding is that credible criminal justice pressures do not diminish the effectiveness of treatment. These programs can do a much better job than imprisonment of reducing subsequent criminal behavior.

Effectiveness will not always mean total abstinence. Drug dependence is characteristically a chronic, relapsing disorder, and more than one episode of treatment typically is required. The better the quality of programs, staff and services, the more likely that an episode of treatment will have a lasting, powerful effect. Money spent to improve and expand treatment is money well spent — not only for those treated, but for everyone whose lives are affected by them.

October 21, 1990

Lawrence S. Lewin, *president of Lewin/ICF, a consulting firm in Washington, D.C., chaired a committee of the Institute of Medicine that studied drug treatment programs.* **Dean R. Gerstein**, *now with the National Opinion Research Center, was the committee's study director.*

★ ★ ★

Making Sense of Statistics in the Courtroom

Stephen E. Fienberg and Miron L. Straf

Some of society's most important court cases are decided on the basis of statistics, often because no substantial direct evidence is available. Yet, too often, statistical analyses are confusing to non-experts. Juries and judges struggle to decide whether to believe the statistician for the plaintiff or for the defense, grappling with such concepts as regression analyses and statistical significance.

One way to ease their task would be to set better guidelines for people who provide the analyses and inferences, mostly statisticians, economists and other social scientists. Many experts now called on to testify in court are caught in a dilemma: They are almost always hired by attorneys for a particular side who want them to testify in a certain way, yet they are expected to maintain a professional commitment to a complete, accurate and balanced inquiry that will guide the court to a wise decision.

This puts the squeeze on experts whose testimony may be crucial to the outcome. Statistical evidence has been central to such celebrated disputes as whether Agent Orange harmed Vietnam veterans and whether housing discrimination in some cities was so pervasive that busing was needed to desegregate the schools. It also has been an essential part of such important cases as these:

- A federal appellate court relied on statistics in overturning an order of the Consumer Product Safety Commission (CPSC) that banned the use of ureaformaldehyde foam insulation in residences and schools. The court said the CPSC used data incorrectly and produced an inaccurate model of the foam's cancer risk to humans.
- A nurse in Maryland was acquitted of murder after a jury rejected the argument that she fatally injected several

patients in a hospital. The state had relied exclusively on statistical models in charging that it was almost impossible for so many patients to die while the nurse was on duty unless she had a hand in their deaths.

Published opinions in federal courts suggest that this use of statistical evidence is expanding dramatically. Statistical methods and analyses are seen as invaluable by many lawyers who use them to present succinct summaries of complex data, to provide a reliable basis for predictions, to prepare quantitative estimates of damages and to clarify complex relations.

Yet this trend also has its costs — and not only the direct costs of hiring expert witnesses or having longer trials. There also is the danger that an increasing dependence on complex statistical procedures could lead some Americans to feel that justice is being turned over to technocrats.

As the use of statistics in judicial proceedings grows, the proper role and conduct of statistical experts need to be clarified. A committee of the National Research Council, which we helped lead, recently suggested some guidelines.

Most simply, we said expert witnesses should never present opinions they do not believe. If they state an alternative viewpoint hypothetically, they should make clear what they are doing. Also, even though a court may not require them to volunteer information, statistical expert witnesses should do so anyway if they believe it is necessary to maintain honesty and accuracy.

Statisticians testifying as experts should be free to carry out their analyses in a scientific manner with access to the necessary data and their colleagues. And if they work under a contingency-fee arrangement in which they are paid only if their side wins the case, this fact should be a matter of court record.

Judges, meanwhile, could make better use of statistics through greater use of court-appointed statistical experts.

Establishing these and similar guidelines would be of great benefit to all of us who might one day find ourselves involved in a lawsuit with expert testimony. As our courts depend on statistics more and more, it is essential that we have confidence in the experts who stand behind them.

February 14, 1989

Stephen E. Fienberg, *dean of humanities and social sciences at Carnegie Mellon University, chaired a committee of the National Research Council that studied the role of statistical evidence in the courts.* **Miron L. Straf**, *director of the Research Council's Committee on National Statistics, was the study director.*

★ ★ ★

The Economics of Reality

Richard H. Thaler

Human beings make mistakes. While this statement may not seem controversial, it flies in the face of much of traditional economic theory. The theory does not match the reality of human behavior, and that is a dilemma at a time when our nation faces economic problems ranging from the federal budget deficit to the trade gap.

Regardless of their political persuasion, most economists build their models on the assumption that investors, managers, workers and consumers are highly rational. To model complex situations, economists assume that people act as if they are able to make economic forecasts with the same sophistication as a computer running an econometric program. This premise of economic rationality has proved successful in some applications in the past, but it is increasingly belied by experimental evidence from those who study the psychology of decision-making, and it is inconsistent with the way economists talk about the behavior of their spouses, colleagues and government leaders.

As a recent National Research Council report on the social and behavioral sciences discussed, people are best thought of as having "bounded rationality," to use the term coined by Nobel laureate Herbert Simon. Humans can only process limited amounts of information, and their memories are unreliable. To cope with their limitations, people adopt simple rules of thumb to solve complex problems. These rules work well in most situations — but not always.

Suppose, for example, you were asked to guess the ratio of homicides to suicides in the United States. If you are like most people, you would judge homicides as more frequent. In fact, there are more suicides. Psychologists Daniel Kahneman and Amos Tversky use this example to illustrate that people often judge frequency by ease of recall. When some events are more heavily publicized than others, they are thought to be more frequent, even if they are not.

Other examples come from news events:

• In the spring of 1986, there was a flurry of terrorist activity in Europe, and several travelers were killed by a bomb placed in the Rome airport. By summer, millions of Americans had canceled their plans for European vacations and took driving vacations in North America instead. Yet the risk of being the victim of a terrorist bomb in Europe was almost certainly lower than the risk of driving on American highways.

• After his poor performance in the Iowa caucuses, many reporters declared George Bush's presidential candidacy dead. Oops.

Both of these are examples of overreaction. In making predictions, people often give too much weight to recent events and too little to long-run tendencies.

The stock market crash last fall provides another example of people making conclusions and acting in ways that defy standard economic models. On Monday, October 19, the Standard and Poor's stock index fell 20 percent. This fall in stock prices, the largest percentage drop in New York Stock Exchange history, followed a fall of 5 percent the previous Friday. A two-day price drop of this magnitude would be predicted by statistical models only once in every trillion years or so.

Following the Monday crash, prices rose 5 percent on Tuesday and 9 percent on Wednesday, the exchange's two largest postwar increases. The following Monday, the market fell 8 percent. It is hard to imagine a model in which such enormous daily price changes could be judged rational in the absence of any important economic news. Rather, investors seemed to be reacting emotionally to the price movements themselves.

Recent research on the stock market has also uncovered numerous "anomalies," facts which are difficult to explain within a totally rational framework. For example, stock prices tend to rise on Fridays and before holidays and fall on Mondays. It seems possible that this pattern is produced by the mood of market participants. Indeed, contrary to most Fridays, prices actually tend to fall on Fridays that occur on the 13th day of the month!

These and other examples all illustrate the basic point that people are human, not economic robots. Economists and the people who use their models — public officials, business leaders and others — must begin to acknowledge and take account of this human side of economic decision-making. A growing body of research is becoming available on how the vagaries of human nature affect economic decision-making, and it needs to be applied in making business decisions, passing laws and determining economic policy.

While traditional economic models *have* made numerous important contributions to our economic understanding, they can be enriched by incorporating human elements of behavior. Stubbornly clinging to them as they now exist is itself irrational.

October 11, 1988

Richard H. Thaler, *the H.J. Louis Professor of Economics at Cornell University's Johnson Graduate School of Management, served on a National Research Council team that studied recent developments in the behavioral and social sciences.*

★ ★ ★

6

SCIENTIFIC HORIZONS

Knowing About Trees

John C. Gordon

Some years ago, a *New Yorker* cartoon showed a well-to-do man and a child walking in the woods. The caption read, "It's good to know about trees; just remember that no one ever made big money knowing about trees."

Times have changed; forests are now important not only economically, but environmentally and politically as well. Nearly every day we hear about global warming, deforestation in tropical areas or a dispute over ancient forests.

What is surprising and disturbing is how little we really know about trees and forests. A report released yesterday by the National Research Council says research on this vital resource is surprisingly weak and fragmented. Our efforts to understand forests are inadequate to sustain the benefits they provide, let alone to meet the needs of a more populous and competitive future.

For example, forests are widely regarded as a safeguard against environmental threats, most recently global warming. That is because trees remove carbon dioxide, a greenhouse gas, from the air through photosynthesis. Globally, trees remove massive amounts of carbon dioxide from the atmosphere. They also shade and cool their local environment by evaporating water from their leaves.

Political leaders and scientists probably are correct to hail forests as environmental buffers and partial antidotes to global

Drawing by Patrick Burke
The Bridgeport Post, Ct.

warming. Yet the knowledge base in which their claims are rooted is frighteningly shallow. Foresters cannot predict quantitatively the effects of forest loss — or gain — on climate. They could not respond with much precision if a political leader asked how much, say, a billion new trees would relieve global warming. Their uncertainty might cause the leader to think twice about acting at all.

Similarly, little is known about the biology of thousands of forest species, particularly "minor" components like shrubs, insects and microbes. This is especially true of tropical forests in the Amazon, Africa and elsewhere, where biological diversity is threatened. Those fighting to preserve these precious areas are hampered by not knowing what they contain.

The same problem applies closer to home. It is shocking how little we know about the biology and biotechnology of the several hundred species of North American trees upon which we depend. Nor do we know much about forests in a system sense. The oldest forest ecosystem studies are about 25 years old — and there are only a few of them.

This lack of knowledge hinders not only environmentalists but those who depend on forests for their livelihood. The worldwide demand for paper, lumber and other forest products has been growing steadily, but timber companies need improved methods to overcome insect, disease and pollution threats. In urban forests, meanwhile, only one in eight trees is being replaced; new techniques could help preserve these groves that so enrich city life.

Across-the-board improvements in forestry research are essential for all these purposes. Our efforts to understand forests must become broader and deeper, making use of the latest scientific techniques. Recombinant DNA technology, for instance, could help reveal the genetic structure and change in forest organisms. This could lead to new commercial trees that resist insects and disease more effectively. Increased understanding of the interaction of plants, animals, microbes and people in forests would enable foresters to manage these complex ecosystems more effectively.

At the precise time that we expect more from forests, however, we actually are *reducing* our efforts to learn about them. Since 1978, the number of undergraduate degrees awarded in forestry and related fields has declined by half. The research budget of the U.S. Forest Service has dropped in buying power even as forestry research conducted by industry has decreased. Many research facilities are outmoded.

This is surely foolish, and sweeping changes are needed to remedy the situation. More competitive research grants, a national forestry research council to provide leadership, and centers of research focused on major areas of concern are all ingredients of a revitalized forest future.

Romantic visions of the forest primeval are fine for story books but inadequate for the environmental and economic challenges we face. Without our urban and rural forests, our cities will be hotter, our countryside windier and drier, and our supply of wild birds and animals smaller and less diverse. We need our forests and must learn more about them if we are to keep and use them.

October 11, 1988

John C. Gordon, *dean and professor of the School of Forestry and Environmental Studies at Yale University, chaired the National Research Council committee that examined forestry research.*

★ ★ ★

Together to Mars — But with Deliberation

Eugene H. Levy

The United States and the Soviet Union are both planning missions to Mars, and a joint project has been suggested as a way to share costs and promote goodwill between the two superpowers.

Even before the recent summit conference, Soviet President Gorbachev suggested the possibility of cooperative Mars exploration. President Bush has committed the United States to manned exploration of Mars, but some question where the money will come from.

Increased cooperation with the Soviets sounds attractive. But is it realistic? How should it be done? And, considering the complexity of a Mars mission and the differences between the two countries' political systems, technologies, languages and the like, how far does it make sense to go at this time?

I chaired a National Research Council committee that was asked by the National Aeronautics and Space Administration (NASA) to define the best approaches to international cooperation in Mars exploration. We concluded that cooperation with the Soviets — along with European countries, Japan, and Canada, among others — could produce many benefits. We recommended a U.S.-Soviet cooperative program of unmanned robotic missions to visit Mars, make measurements, collect samples and, leaving instruments behind, return to Earth.

However, we also found that close technical cooperation with the Soviets would be likely to create problems at this stage of the relationship. For now, we recommended against the two nations dividing responsibilities for actual joint missions.

Mars exploration could serve many human interests, spanning national boundaries. The surface of Mars reveals a history of startling change. Cold, dry and nearly airless today, Mars apparently once had abundant flowing water, suggesting higher temperatures and a thicker atmosphere than now. The planetary changes recorded in Mars' surface may represent a spontaneous transition from a potentially life-supporting environment to a lifeless one.

We do not yet understand what causes such great change in a planet's environment. One motivation for studying Mars is to find out. Learning why Mars changed will teach us about planets in general, something essential for understanding Earth. We will not be able to understand and predict Earth's behavior confidently until our theories also explain what happened to Mars.

For these and other reasons, Mars exploration is a natural candidate for ambitious international cooperation. In such an effort, the United States and the Soviet Union would occupy unique positions. Only they have the experience, the current technical capability and the expressed vision to undertake or lead an intensive Mars exploration. For some time, the overall approach to Mars exploration will depend on commitments made by the two superpowers.

However, the United States and the Soviet Union have little experience with technical cooperation. Communication

> Our best evidence, from the 1976 Viking mission, indicates that there is now no life on Mars. But wetter and warmer conditions in the past may have enabled life to begin and evolve. Scientists speculate that biochemical or structural fossils might have been left by early martian life.
>
> The space age gave human beings the vision of Earth as a planet and impressed upon us the isolated and fragile nature of Earth's environment. Our planetary perspective was further sharpened after investigations of Mars revealed that planets can undergo extraordinary environmental change, perhaps even from a life-supporting world to a lifeless one. That prospect was not seriously considered before explorations of Mars. Studying our planetary neighbor provides an irreplaceable perspective on planetary environmental change.
>
> —E.H.L.

is poor; neither side fully understands the institutions, practices or capabilities of the other. Close cooperation would raise concerns — some reasonable, others unreasonable — about technology transfer. Moreover, space exploration projects involve long-term commitments. Joint missions, in which each nation is greatly dependent on the other, could become hostage to unexpected political events.

Given these problems, it makes sense for the two countries to develop cooperation gradually, although steadily, beginning with robotic missions that use artificial intelligence to explore Mars and return samples to Earth. Automated exploration would be extremely beneficial in itself because of its scientific value, social inspiration, potential to spur technology and relative economy. Decisions about human exploration could be made later.

Both nations would benefit technologically from robotic exploration of Mars. Important advances could be expected in rocketry, robotic manipulation, machine intelligence and other fields. It is not clear that either nation would want to relinquish to the other the development of the major enabling technologies.

An initial phase of Mars exploration would involve about half a dozen of these robotic excursions. The United States and the Soviet Union could jointly lead the world in such a project. Each would conduct three coordinated excursions. Planning and scientific investigations could be carried out in close cooperation. There would be ample opportunity for other nations to participate.

This approach, increasing cooperation substantially but stopping short of fully joint missions for now, would allow the two superpowers to begin immediately a project of historic importance on behalf of the whole human race. Both nations would achieve economies and would accomplish scientific objectives of global importance. They also would begin to build the experience, knowledge and trust that could foster even closer cooperation in the future.

June 10, 1990

Eugene H. Levy *is head of the planetary sciences department at the University of Arizona, Tucson.*

* * *

The Less-Noticed Worldwide Revolution

Peter H. Raven

Many parents across the country reportedly have been stunned to discover that their teenagers are baffled about the significance of events in Eastern Europe and the Soviet Union over the past several months. How, these parents wonder, can young people be so ignorant about such historic occurrences?

I share their concern, but I think many parents are as culpable as their children when it comes to awareness of another, equally important upheaval taking place in the world today. I refer to the revolution in biology, which is likely to change the course of human history as profoundly as anything in today's political arena.

The recent transformation of the biological sciences is comparable to the tumult that occurred in physics earlier this century with Albert Einstein and others or in astronomy at the time of Galileo. When one considers the impact those two earlier upheavals had on subsequent history — on politics, warfare and other events far beyond the world of science — it is not an overstatement to describe current advances in biology as among the most consequential developments of modern times.

During the past two decades, biology has been transformed from a collection of single-discipline endeavors to an interactive science of extraordinary vitality. Starting with the establishment of the structure of DNA and continuing with the demonstration that genes could be modified and moved from one organism to another, the flow of biological discovery has swelled from a trickle to a torrent. Advances have followed rapidly from new methodologies, such as the use of recombinant DNA, monoclonal antibodies, microchemical instrumentation and computers.

A complete understanding of living systems at the molecular level now seems to be at hand. As a team of more than 100 scientists that I led for the National Research Council concluded recently, modern biology is poised to make fundamental discoveries critical to understanding how humans resist infection, how a fertilized egg develops, and how humans dream, imagine and reason. Researchers are revealing the mechanisms underlying simple forms of learning and short-term memory. They are using sophisticated molecular and genetic techniques to analyze genetic differences between species, clarifying how life on Earth emerged.

Many of these secrets of nature are so fundamental that uncovering them could affect our human self-image as deeply as did Copernicus' revelation five centuries ago that the Earth revolved around the Sun, rather than vice versa.

The "new biology" clearly is of immense importance to the U.S. economy. Technologies based on biology will provide the basis for advances not only in predictable fields such as pharmaceuticals, but in a wide range of manufacturing and service industries. Companies throughout the United States, Europe and Japan have begun making major investments in new products based on genetic engineering, fermentation and other biological applications. Our country will have no choice but to compete vigorously in this new arena.

In health care, new biological information promises to make possible significant advances through new therapeutic drugs and improved methods of diagnosis for AIDS, cancer, Alzheimer's disease and other ailments. In agriculture, farmers will control pests increasingly through biological techniques rather than with chemicals, increasing their profits while providing healthier food and a safer environment.

At such a historic juncture, it is ironic that the United States faces a projected shortage of biology researchers and trained technical experts and that many U.S. universities and research centers require better facilities and instrumentation to carry out world-class research. The future of biology also is clouded by the rapid extinction of plant and animal species occurring around the world as human populations expand and natural habitats are destroyed. Each form of life that disappears takes with it a pattern of gene expression that evolved over millions of years, destroying a piece of the foundation upon which all of biology is built. This priceless worldwide genetic heritage must be protected.

It would be a tragic squandering of opportunity if inadequate resources and the destruction of species were allowed to slow the pace of continued biological discovery. Just as recent events in the communist world have reshaped the world's political landscape, the biological revolution has the capacity to transform our lives in extraordinary, if unpredictable, ways. It is a revolution whose reverberations will be felt even by people who now are oblivious to its occurrence.

April 29, 1990

Peter H. Raven *is director of the Missouri Botanical Garden.*

★ ★ ★

Searching for Buried Treasure

Charles A. Bookman

In an era when our nation has sent explorers to the moon, probed the planets and is considering a manned voyage to Mars, another frontier much closer to home lies largely unexplored. This prize, containing billions of dollars in resources, is our national seabed. The United States has barely begun to capitalize on the plants, animals, minerals, recreational opportunities and other riches lying off its shores.

In 1983, President Reagan established the 200-mile zone beyond the coastline that is now under our national jurisdiction. However, this 3.9 billion acres of ocean territory has yet to be explored systematically. No single federal body coordinates the often conflicting interests of private companies, universities, the military, states, cities, federal agencies and others who have an interest in this natural underwater treasure chest.

As an expert committee of the National Research Council urged in a report recently, the federal government needs to develop a coherent national policy to analyze the seabed more thoroughly as a first step to putting it to better use.

Ocean businesses already contribute 1.7 percent to our country's gross national product. In 1987, this amounted to $76 billion, mostly from fishing and offshore oil and gas drilling. The importance of fishing is apparent to consumers, and the substantial U.S. dependence on offshore drilling is certain to expand as onshore resources are exhausted.

Yet the seabed offers more than fish and fuel; it also is a trove of minerals. Off Alaska, miners extract gold from the ocean floor. In the central Pacific off Hawaii, potato-sized nodules rich in strategically important minerals await harvesting. Within the continental United States, offshore sand and gravel deposits are occasionally tapped to control beach erosion or for other construction purposes. While most ocean minerals will not be mined until prices improve, the United States would benefit immediately from a clearer picture of its holdings of critical minerals such as cadmium and cobalt, for which it now is overly dependent on a few foreign countries.

The oceans also are essential to the world's communications systems. With the rapid development of fiber-optic technology, underwater cables have replaced satellites as the transoceanic medium of choice. Half of all overseas calls are now transmitted through cables. Our armed forces rely on the oceans as an operational area for submarines, monitoring and listening systems, and anti-submarine and mine warfare.

Other ways of using the seabed have been tried or suggested, as well. More than 750,000 tourists have ridden to the sea floor during the past few years in specially designed submarines, mainly in shallow water at tourist resorts in the Virgin Islands, Hawaii and other places. It does not take much imagination to envision excursions to the growing number of marine sanctuaries and memorials, such as the sites of the *Monitor* and the *Titanic*.

Our oceans also could be used for farming and for energy systems. Hawaiian farmers already are using nutrient-rich seawater to help grow giant strawberries and other products. Energy innovators are studying ways of harnessing the heat differential between land and sea to generate electricity or of using cold seawater to produce air conditioning for seaside buildings. Still another possibility is to use the seabed as a site for unwanted urban wastes. Sewage sludge might be placed beneath the seabed and isolated with clean sand.

Ideas such as this last one require considerable study to guard against environmental or other problems, but the general concept of using our oceans more imaginatively clearly deserves closer consideration. President Bush and Congress have an opportunity to launch a major exploration of the nation's last frontier, much as President Thomas Jefferson authorized the Lewis and Clark expedition to investigate the American West nearly two centuries ago. Such a mission would expedite exploration of the seabed, head off potential conflicts among users and perhaps even capture the imagination of the public. It would be most valuable if a special commission also were established to determine priorities for seabed exploration and development and to generate needed technology.

As the Research Council committee pointed out, our seabed is a national treasure of unprecedented dimensions, one whose resources are barely tapped. It will require considerable cost and long lead times to make the most of this bounty, but the benefits to the nation will be substantial.

March 4, 1990

Charles A. Bookman *is the director of the Marine Board of the National Research Council.*

★ ★ ★

The Energy Crisis Beyond the Persian Gulf

David L. Morrison

Fifty years from now, when Saddam Hussein is just a bad memory, Americans could face an even worse predicament. This potential danger will be affected directly by actions we take in response to the Persian Gulf crisis.

This longer-term threat is not military but environmental: global warming and climatic change. Scientists are still studying the likelihood and possible extent of such changes, but warming of the Earth could have a devastating impact on the world of the future. Some plausible scenarios foresee a rise in sea level, reduced agricultural productivity and millions of environmental refugees.

During the past five years, with gas prices so cheap, Americans showed little interest in energy policy. Now policymakers are scrambling to reduce oil imports and promote use of domestic energy sources. Additionally, we must seek to reduce our consumption of fossil fuels to ease the threat of global warming. Both goals can be achieved by moving faster to improve our nation's energy productivity and to adopt alternatives such as biomass fuels and solar-powered photovoltaics, recycling technologies, more efficient buildings and transportation systems, and better methods of storing electricity.

These and other alternatives have languished over the past decade. During the 1980s, federal funding of the Department of Energy's (DOE) civilian research and development program on solar and renewable resources plunged by nearly 90 percent. DOE research on energy conservation declined by 61 percent, and research by private companies also fell.

Although the nation cannot invest in every alternative, some deserve closer consideration. At DOE's request, a committee of the National Research Council, which I chaired, recently identified several that look especially promising on both environmental and economic grounds.

For example, cars and light trucks are a major source of the excess carbon dioxide in our atmosphere. The average fuel efficiency of new cars in the United States is only 28 miles per gallon. With stringent government policies and a greater effort by industry, this probably could be raised to 45 miles per gallon within a decade. Another goal should be to improve batteries to provide motorists with electric cars that offer better performance at lower costs without the emissions.

Energy use and greenhouse gas emissions in our residential and commercial buildings can be slashed by more than 70 percent with improved materials for walls, windows and roofs; more efficient heating systems, air conditioners and lights; and other strategies.

Research also should be focused on coal-fired generators, which produce more than half of the nation's electric power and most of its acid rain. With existing technology, sulfur dioxide emissions can be reduced to ease acid rain — but at the price of lower efficiency. As a result, more plants are needed and more carbon dioxide may be produced. Rather than pursuing tradeoffs between two undesirable kinds of emissions — sulfur dioxide and carbon dioxide — we should find ways to reduce both while increasing efficiency.

Renewables also deserve more attention. The price of

electricity produced by photovoltaic modules has plummeted since the early 1970s but still exceeds the cost of power from public utilities. Manufacturing research could bring down the price of such renewable technologies.

If budget constraints prevent increased funding for this kind of research, DOE should consider scaling back some magnetic fusion research. Commercially viable fusion reactors are highly unlikely to make any significant additions to the U.S. electricity generation mix before the year 2050. Efforts in this field should shift towards more basic research and greater international collaboration. This would free up funds for research that promises a more immediate and lasting energy payoff.

Obviously, research alone cannot ensure energy security or protect the environment. Changes in public policies, such as new tax credits or stricter energy efficiency standards, should be considered as well. In addition, the government might stimulate consumers to adopt more efficient home furnaces and other underused technologies.

DOE now spends $2.2 billion annually on civilian energy research and development, and it may well spend more before the dust settles in the Persian Gulf. This effort should be applauded — and expanded. It also should be crafted in a way that will serve us well, not only through today's crisis but in the uncertain climate of the future.

September 16, 1990

David L. Morrison *is technical director of the energy, resource and environmental systems division at MITRE Corp., in McLean, Va.*

★ ★ ★

The Challenge to Human Uniqueness

Herbert A. Simon

The rapid development of artificial intelligence in computers is about to challenge our sense of human uniqueness as profoundly as anything since the days of Copernicus or Darwin.

At one time, we might remember, human beings thought they had been placed in the geometric center of the universe. Then Copernicus came along and said we humans had it all wrong, that we really live on a planet circulating around the sun. So mankind had to develop a new sense of its uniqueness that no longer relied on being physically at the center of things.

Next came Darwin. He pointed out that we had been resting our notions of uniqueness on the idea that we are a specially created species unlike any other. Darwin showed that the human species evolved through processes of mutation and selection just like all the others. So now mankind had to give up its notion of uniqueness not only in the universe, but among species.

Nevertheless, we humans have continued to think of ourselves as unique in the years since Copernicus and Darwin. Why? In considerable part because of our capacity to think and reason. Other animals also think, of course, but we are seemingly the only ones who can think complex thoughts, abstract thoughts, thoughts involving the use of language.

Developments in artificial intelligence, the study of computers doing intelligent things, are now challenging this aspect of uniqueness.

What do we mean by intelligence? How do I know that a person is thinking? I can check whether the person has a studious frown on his face as he ponders a problem, of course, but that is not very reliable evidence. The only empirical way to decide is to give a person a task and then judge on the basis of his or her performance whether a thought has taken place in reaching a solution.

It is only human chauvinism to refuse to call something non-human, such as a computer, intelligent if it does the same. Computers today now play chess just below the grandmaster level. They can examine data and discover scientific laws. A program called BACON that I helped develop was given the distances of the planets from the Sun and their periods of revolution. It discovered in less than a minute that the periods vary as 3/2 powers of the distances. This is Kepler's Third Law, an important discovery of the 17th century. Other computer programs are diagnosing medical illnesses, prospecting for ore and synthesizing chemical reactions.

True, computers have not yet been able to write good poetry or great music, or to solve certain kinds of analytic problems. Large areas of human thought processes still have not been explored. Nonetheless, having worked with artificial intelligence for almost three decades, my bet is that every kind of human thinking will eventually be able to be performed by non-human systems.

It is significant in this regard that robotic devices for use in variable physical environments have proven much more difficult to develop than computers that mimic abstract human thought. One reason is that higher human cognitive skills have been evolving for only a couple of million years, whereas our sensory and motor skills evolved over 400 million years and therefore are more sophisticated and harder to replicate. Given how proud we are of our intelligence, it should give us pause to remember that it's easier to automate a professor than a bulldozer driver.

What are the likely implications for human society of these developments in artificial intelligence?

One of the most encouraging possibilities concerns education. Much teaching today is inefficient, largely because it is based on remarkably little fundamental understanding about how a student's brain processes knowledge. New insights from artificial intelligence and related fields may enable us to revolutionize education, much as medicine was transformed when researchers finally began to understand the biological bases of disease.

Increased knowledge about ourselves also should help us to become better problem-solvers and decision-makers. The threat of nuclear war, stress on the environment, scarcity of resources and other problems in the world today are caused ultimately not by technology, but by ourselves. We will solve these problems only when we learn to improve the use we make of our own minds.

Most important is how we will change our image of ourselves and our sense of place in the universe. It is important to recall that most people did reconcile themselves to the discoveries of Copernicus and Darwin, and did not feel any the worse for it. One can have confidence that people in the future also will find a way to describe their place in the world without having to believe that they are unique as thinkers.

Mankind's development of a new self-concept is likely to be as valuable as any specific benefits that it will gain from computers themselves. It is ironic, perhaps, but the ultimate benefit of our search for smarter machines may well prove to be this deeper knowledge of our own thinking and of ourselves.

June 23, 1985

Herbert A. Simon, *winner of the Nobel Memorial Prize in Economic Science, is a professor of computer science and psychology at Carnegie Mellon University.*

★ ★ ★

Making a Map of the Human Chromosomes

Bruce M. Alberts

Imagine you are an ambulance driver and someone's life depends on your finding an address — but you have no map. That's the dilemma faced by medical researchers, who must search for the cures to serious genetic diseases without knowing where the relevant genes are located on the human chromosomes.

The result is that research today proceeds much more slowly than it might on heart disease, cancer, certain kinds of Alzheimer's disease, cystic fibrosis and some 3,000 other disorders with a genetic component. Researchers must spend vast amounts of their time and resources searching again and again for genetic needles in the haystack of the human chromosomes; a typical set of chromosomes holds about 100,000 genes, of which fewer than 1,500 have been charted.

Like the ambulance driver searching blindly along city streets, these researchers could do their work far more effectively — and save many more lives — if they had a decent road map. Fortunately, that is now possible. An expert committee of the National Research Council, which I chaired, concluded recently that it has become feasible to map all the genetic material in the human chromosome, which scientists refer to as "the human genome."

Our committee, which included two Nobel laureates and a number of the world's leading biologists, urged the federal government to launch a program to create such a genetic map. The project, the largest research effort with a defined focus ever undertaken in biology, would greatly increase our understanding of the human organism and promote rapid progress in controlling many diseases.

The human body is extraordinarily complex, containing about 10,000 billion individual cells. Each cell carries a complete set of genetic blueprints for the entire body. These blueprints are stored as DNA, a molecule that when highly magnified looks like a very long and thin twisted rope ladder. A single DNA molecule forms each human chromosome, containing the genes that determine everything from the color of a child's eyes to the likelihood of an adult's suffering from certain forms of manic depression.

The goal of mapping the human genome is to identify the location of all the genes on each chromosome, which is to say on the DNA molecules. A genetic map would not cure genetic diseases by itself; it would only be a tool to help researchers analyze these diseases in their laboratories. Yet its value would be incalculable, like trading in the maps of Columbus for detailed satellite photos. Locating desired genes is essential because it leads directly to the discovery of the molecules that cause genetic diseases.

Scientific expertise in gene mapping is still in its adolescence, although major advances have been made in the past few years. Therefore, a major priority in the near future must be to develop more powerful techniques. Our committee decided that the best way to accomplish this would be to divide the task among competing research teams around the country, with each team using its own techniques to map a large amount of DNA. The techniques could then be compared and the best ones selected for further use.

We estimated that several types of detailed maps of the entire human genome could be completed within five to ten years. However, the most detailed description is much more difficult to obtain. It would give the chemical composition of all the chromosomes, showing the exact order of the subunits, or nucleotides. There are about three billion of these subunits in the human genome, arranged in a precise order like beads in a necklace. A technique called "DNA sequencing" allows this order to be determined.

One way to understand the difference between mapping and sequencing is to think of a giant library. Mapping is like noting the titles and order of all the books on the shelves; sequencing means actually reading the books.

With present techniques, it would take an estimated 30,000 "person years" of labor to obtain all of this sequencing information. Rather than attempt such a costly project now, our committee recommended the immediate support of many competing efforts to improve sequencing methods and lay the groundwork for a full-scale sequencing effort.

We urged that new funds be provided for a special "human genome project," costing about $200 million annually for 15 years and including both mapping and sequencing. The funds would go primarily to individual investigators and small centers, with scientific oversight by an independent board of leading scientists. A central data bank and stock center also would be needed to gather and share all the information among the research teams.

The cost of this project would be about 3 percent of current federal spending on basic biology research. Our committee considered that to be a price well worth paying. Mapping and sequencing the human genome would create a tool of fundamental value, providing researchers with the "dictionary" they need to speed future medical advances.

March 20, 1988

Bruce M. Alberts *is professor of biochemistry and biophysics at the University of California, San Francisco.*

★ ★ ★

Developing New Contraceptive Options

Luigi Mastroianni, Jr.

Birth control pills, condoms, intrauterine devices (IUDs), diaphragms, contraceptive sponges, foams and other vaginal contraceptives, and natural birth control methods are the options available to couples in the United States who wish to practice contraception.

These choices are inadequate. In some European countries, couples also can choose among contraceptive implants; injectable contraceptives; and a variety of pills, IUDs and sterilization techniques not offered in the United States. For example, an implant placed under a woman's skin that releases progestin has been available in Europe, Asia and Latin America for the past decade but only recently has received serious consideration in the United States.

An expert committee of the National Research Council and the Institute of Medicine reported this past week that U.S. consumers not only cannot obtain useful contraceptives available in other parts of the world but also may miss out on technologies still in development. These include a contraceptive vaccine, improved techniques of reversible male and female sterilization, a once-a-month pill that induces menses, and methods that interfere with sperm production.

Since the introduction of the pill and the IUD in the early 1960s, no fundamentally new contraceptive methods have been approved for use in the United States. Only one large U.S. pharmaceutical company still maintains a significant contraceptive research program, although some smaller firms and non-profit organizations have stepped up their research efforts.

One might attribute this slow progress to a lack of interest among potential users. However, of the 54 million U.S. women between the ages of 15 and 44 who have had intercourse, 95 percent have used contraception at some time. Many of them — and their male partners — are dissatisfied with the choices obtainable from physicians and pharmacies. Contraceptive failure is common, and unwanted pregnancies lead often to abortion.

Both women and men need new methods to meet their contraceptive needs as they pass through the stages of their reproductive lives. A given method may be most appropriate for young people and for those having intercourse only occasionally, while another method may be better suited to mothers who are breastfeeding and want to space their pregnancies. An increase in the number and type of contraceptive options also would ease important social problems, such as teenage pregnancy, abortion and the spread of sexually transmitted diseases.

The current situation results from many causes, but our committee identified two ways of easing it considerably. First, the Food and Drug Administration (FDA) should adopt a more realistic method of evaluating new contraceptives. For most other drugs and devices, the FDA weighs the risks and benefits for a specific group of users, such as patients with cancer or diabetes. When it comes to contraceptives, however, the FDA assesses the potential impact on *healthy* users rather than considering the special needs of nursing mothers, older women who smoke, and other inadequately served groups. More weight should be given to these variations among potential users.

The FDA also should strengthen its current emphasis on safety by paying greater attention to the effectiveness and convenience of different methods. Methods such as spermicidal jellies and foams are not necessarily safer if they have higher failure rates, especially given the potentially serious health consequences of unwanted pregnancies in some circumstances. Modifying the approval process could make more products available while maintaining rigorous safety standards.

Our other main recommendation dealt with liability laws. Recent product-liability litigation and rising insurance rates have been a major obstacle to contraceptive development. Many pharmaceutical companies view the current situation as so unpredictable that they are unwilling to make costly investments in new products. Our committee concluded that the rules could be changed to ease their concerns without increasing the risks to consumers.

At the least, Congress should enact a statute that gives companies some legal protection when they follow FDA guidelines faithfully in producing a product. If it turns out later that unforeseen and unforeseeable complications occur, the companies should not be held responsible unless it is shown that they willfully withheld relevant information.

These two steps would go a long way toward meeting the public's demand for safer, more effective, more convenient and affordable contraceptives. Unless action is taken to change public policy, contraceptive choices in the United

States in the next century will not be appreciably different from what they are today.

February 18, 1990

Luigi Mastroianni, Jr., *director of the division of human reproduction at the Hospital of the University of Pennsylvania, headed a committee of the National Research Council and the Institute of Medicine that studied contraceptive development.*

★ ★ ★

Farewell to the Night Sky

David L. Crawford

A priceless part of our human heritage is fading into the night sky.

Most Americans are growing up unable to see the stars their grandparents knew so well. They see the night sky only in pictures or at planetariums. This is true not only in cities but also in many suburbs where street lamps and other sources of "light pollution" have obscured our view of constellations, meteor showers and planets.

Indeed, many youngsters may now say, after viewing the night sky in a rural area for the first time, that "it looks just like the planetarium."

Light pollution is not a matter of life and death. Yet it is important nonetheless, profoundly so. We human beings lose something of ourselves when we can no longer look up and see our place in the universe. It is like never again hearing the laughter of children; we give up part of what we are.

Such a loss might be acceptable if light pollution were the inevitable price of progress. But it's not; most sky glow, as scientists call it, is unnecessary. The light that obscures our view of the night sky comes mainly from inefficient lighting sources that do little to increase nighttime safety, utility or security. It produces only glare and clutter, costing more than a billion dollars annually in wasted energy in the United States alone.

For science, the impact has been even more tangible and adverse. Astronomers require observations of extremely faint objects that can be made only with large telescopes located at sites free of air pollution and urban sky glow. For example, scientists interested in how the universe was formed may study the light of galaxies and quasars located incredibly vast distances from Earth. These images offer information about faraway corners of the universe, helping us understand how our own world was formed. Yet, after traveling countless light years, the light from these objects can be lost at the end of its journey in the glare of our own sky.

Space-based telescopes, such as the Hubble Space Telescope, offer one way around the problem. However, large telescopes here on Earth will always be used, if only because they are accessible, cost much less than orbiting devices and can do many jobs more cheaply.

In fact, our experience over the past two decades has shown that space-based astronomy, far from reducing the need for ground-based observations, actually *increases* the demand for these facilities. New telescopes now planned or under construction here on Earth will complement the knowledge we gain from telescopes in space — but only if they are not compromised by encroaching light pollution, as has occurred near Mount Wilson, near Los Angeles and several other older observatories.

Reducing light pollution is not difficult, but it does require that public officials and ordinary citizens be aware of the problem and act to counter it. Low-pressure sodium lights, for example, can replace existing fixtures for most streets, parking lots and other locations. They reduce glare and save money.

Another fairly painless way to reduce light pollution is with outdoor lighting control ordinances, over 50 of which have been enacted throughout Arizona and in several key cities and counties in California and Hawaii. These measures typically require communities to prohibit inefficient, low-quality lighting. Not only do they help preserve dark skies, but they also enhance energy efficiency. An outdoor lighting system recently installed at a prison in Arizona, for example, improved security and reduced light pollution while cutting energy costs by 50 percent. There is no reason that all communities should not have such efficient lighting.

On an individual level, people can help reduce sky glow by using night lighting only when necessary, choosing well-shielded light fixtures, and turning off lights when they are not needed.

Curing light pollution saves money while reducing glare. Unlike other issues involving pollution, it presents us with a rare case where we should strive to be "kept in the dark." The stars above us are a priceless heritage — not only for scientific knowledge, but also for our identity as human beings.

More of our children — and their children — should be able to look up at night and see that the Milky Way isn't only a candy bar.

December 3, 1989

David L. Crawford, *an astronomer at the Kitt Peak National Observatory in Tucson, is executive director of the International Dark-Sky Association.*

★ ★ ★

Setting Our Science Priorities in Order

Frank Press

Science and technology present the Bush administration with some of its best opportunities to leave its mark on history. The super collider, a program to map the human genome, the space station, a new AIDS initiative, and increased research on environmental problems and superconductivity are among the many possible initiatives that could change our world profoundly.

Yet, even as investments in science and technology offer greater promise than ever before for producing significant benefits in health, the environment and other fields, the United States faces unprecedented budget deficits. How, then, is it to pursue these new opportunities while also providing adequate support to smaller-scale research, science education and other activities that are less visible but of equal, if not greater, importance?

That is a dilemma facing the new president and Congress, and it is made more difficult by the inadequate system now in place to make federal budget decisions about science and technology. Although effective in the past at helping the United States assume world leadership in these fields, the system is unable to provide us with clear national priorities in the face of these historic opportunities and constraints.

The system does do a good job of setting priorities within specific agencies involved in science, but not when it comes to looking across agency lines and establishing priorities overall. Both the executive branch and Congress are left focusing on the trees instead of the forest.

For example, when researchers in Zurich announced in late 1986 that they had discovered materials that become superconductive at much higher temperatures than anything recorded previously, they set off an international race to develop new applications and industries. Officials in Washington soon began asking what the United States was doing in the field, only to discover that the federal effort was split among five agencies with no capacity for overall assessment in place. In the end, Congress had to create a special commission to provide direction.

Similarly, federal efforts to understand global warming and other kinds of climate change are now divided among the Environmental Protection Agency, the Department of Energy, the National Science Foundation, the National

Aeronautics and Space Administration, the Department of Agriculture and the National Oceanic and Atmospheric Administration, among other agencies. Such a multiplicity of efforts has many benefits, but it should not be as difficult as it is to find out what the government is doing overall about climate change.

The federal government now spends more than $60 billion annually on science and technology activities, and it needs to allocate the money more effectively. At congressional request, the National Academy of Sciences, the National Academy of Engineering and the Institute of Medicine recently offered some suggestions on how this might be accomplished.

Our basic message to both President-elect Bush and the Congress was the same: Set clearer priorities before dividing the budget pot among the different agencies. Specifically, we said the president should establish overall goals in science and technology that individual agencies can use as guidelines in preparing their own budgets. Congress should follow a similar process.

These goals should be set not only along traditional agency lines, but also in terms of how they will contribute to the nation's underlying science and technology base — its work force and research facilities — or to broad national objectives, such as industrial competitiveness and environmental protection. Major initiatives such as the space station may need to be considered as a separate category.

A greater effort should also be made to distinguish between military and civilian research in the budget. Much military research has limited application to the civilian sector, and lumping the two together tends to overstate the true size of the U.S. science and technology enterprise. Reforms like these do not alter the traditional prerogatives of government officials. Nor do they lead to a centralized science bureaucracy, which might threaten the flow of unconventional ideas that are so essential to the scientific process.

Instead, rationalizing the budget process in this way will help officials see the "big picture" on questions as vital as AIDS, the space program, the global environment and agriculture. They will become better able to put science and technology to work to solve the problems that lie ahead for our nation not only over the next four years, but in the decades to come. Science, of all pursuits, ought to be handled more rationally.

January 10, 1989

Frank Press *is president of the National Academy of Sciences.*

7

INTERNATIONAL AFFAIRS

Getting Even in International Technology

H. Guyford Stever

"Don't get mad. Get even."

Most Americans probably identify those words more with bumper stickers or bruised athletes than with such genteel subjects as scientific research or industrial production. But "getting even" — not for revenge, but to catch up — is precisely what American engineers and technologists need to start doing with their counterparts in other industrial countries. They have to pay as much attention to scientific and technological developments abroad as foreign experts do to developments here.

For many years now, technical experts from Japan and other countries have devoted considerable resources to monitoring U.S. universities, research laboratories and other facilities to learn what our scientists and engineers are doing. The result often has been that these experts return home and apply their findings to profitable new products and processes. They use our knowledge base for their own advantage. This is not unfair; it's competition.

It does no good to accuse foreigners of "ripping off" American expertise. There is no way to prevent other nations from monitoring us even if we wanted to, which we don't, since almost any action would inhibit the flow of information among our own researchers as well. For example, one cannot simply prevent all foreigners from reading U.S. technical

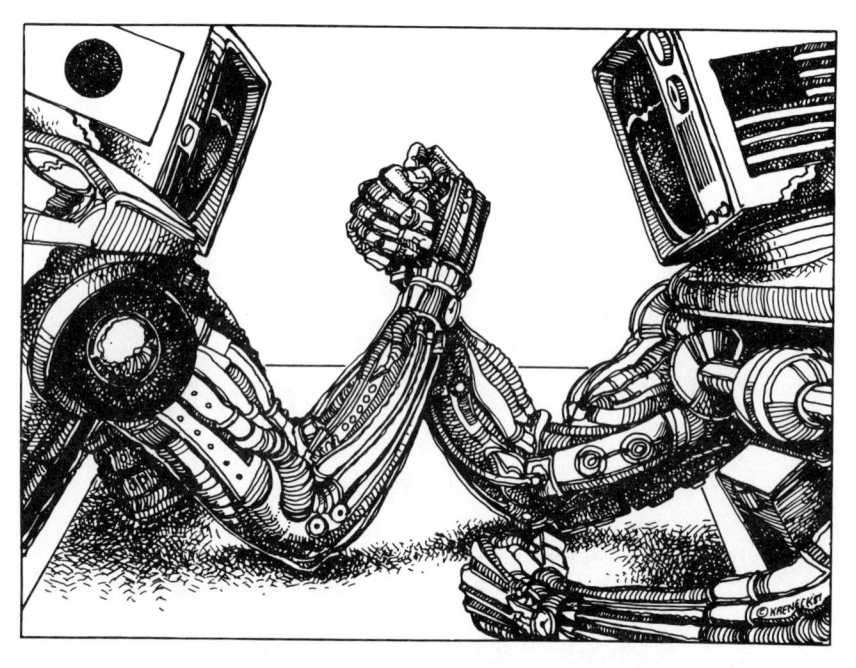

Drawing by Kevin Kreneck
Roanoke Times & World-News, Va.

journals. Such technological protectionism is unrealistic; technology diffuses inevitably and quickly.

Rather than remaining frustrated about the situation, U.S. engineers and companies should be responding in kind, overcoming traditional isolationism and monitoring foreign research more vigorously.

We have a lot to learn from our foreign colleagues. Israel and Japan are active in artificial intelligence research. Denmark and the Soviet Union are developing new kinds of cements. West Germany and Hungary are among the top nations in developing manufacturing systems. Taiwan and Switzerland have been involved in the recent advances in superconductivity. Austria, Germany and Japan are developing new engine designs. Almost two-thirds of all research publications in engineering and technology now originate outside the United States.

Yet, despite this activity, many U.S. engineers make little effort to learn about foreign research, maintaining their post-

World War II sense of technical superiority. American students demonstrate a similar bias; in the 1984–85 academic year, about 13,000 Japanese students studied in American universities while only 700 American students studied in Japanese universities. While the imbalance with European universities is not as extreme, we still neglect the many fine institutions there.

An expert committee of the National Academy of Engineering, which I chaired, warned recently that this "technological isolation will surely undermine the future of our industries and educational institutions." One danger sign is that, in 1986, the United States imported more high-tech products than it exported.

Part of the long-term solution is for American engineers and companies to become more "worldly," to read about foreign countries, travel more, learn foreign languages and develop close ties with foreign colleagues. Admittedly, this takes time. Yet there are several ways the U.S. engineering community can begin remedying the situation right now:

- U.S. engineering schools should instill in their students a greater appreciation for foreign achievements and provide more opportunities to study abroad. One good example is Stanford University's arrangement with Kyoto University for U.S. students to study in Japan. Universities in our country especially need to help their students gain spoken and technical competency in Japanese and other Asian languages.
- Engineering and professional societies should seek to increase the participation of their U.S. members in international activities and to help them learn about foreign technological developments.
- U.S. companies involved in engineering and technology should develop closer working relationships with foreign universities, research institutes and companies that are technologically advanced in relevant fields.
- Finally, the federal government, through the National Science Foundation, should increase its funding for U.S. participation in international collaborative engineering research and education. NSF also should require its sponsored re-

searchers to demonstrate that they are aware of engineering knowledge generated abroad.

In these and other ways, our country's engineering community must change to reflect the fact that the technical world has become more international and interdependent. Otherwise, it will be left behind in its own ethnocentric dust — and our country will suffer as a result. It's time we started getting even.

September 20, 1987

H. Guyford Stever *was science adviser to President Ford.*

★ ★ ★

The Growing International Competition for Brain Power

Peter W. Likins

The United States has a problem: Hong Kong wants John Chen.

Chen is not an international criminal or political activist. He is the former chairman of the chemical engineering department at Lehigh University. He left China when he was seven years old, arriving in the United States at the end of World War II. He excelled as a student, earned a doctorate and reached the top echelon of his profession.

Now, the Chinese want him back. The Hong Kong University of Science and Technology will admit its first students in 1991, and it is urging Dr. Chen and others like him to move to Hong Kong to become deans and recruit a faculty for the new campus. The university is expected to hire 1,000 professors over the next decade.

Not all these professors will come from our shores, but the United States is the mother lode of the world's technological talent, and half of its engineering professors under age 36 are foreign born. South Korea, Taiwan, Singapore and India also have mounted aggressive campaigns to attract these experts back to their native countries.

The United States will be the big loser if they succeed. Rarely seen by the general public, and unknown even to many people on their campuses, engineering professors are a critical national resource. They carry out much of the research and train the talent that keep our country on the cutting edge in computers, robotics, aircraft design and other technologies vital to our future.

Twenty percent of the engineering faculty at U.S. universities is expected to retire within five years. However, there are ominously few U.S. students prepared to take their places. Since 1981, more than half of the engineering Ph.D.s in our country have been awarded to foreign-born students. These experts are the ones now being wooed by their native lands. If the United States is to maintain its strength in technology, it must continue to retain many of them as professors.

Even more important, our country ought to be doing a better job of training its own sons and daughters for these careers. Our elementary and secondary schools need to excite young Americans in science and math and prepare them better for higher studies. The National Assessment of Educational Progress reports that only 7 percent of 17 year olds still in school are ready for college science and engineering courses.

Once in college, students need greater incentives to pursue advanced technical degrees. The federal government offers engineering graduate students substantially less fellowship support than it did 20 years ago. It should not only expand this support but also forgive some loans for engineering graduates who teach for a few years. More also needs to be done to recruit women and minorities, both of whom are underrepresented in engineering and could contribute much to our nation's technological and economic success.

In addition, those of us within higher education have to do a better selling job. Low starting pay and the arduous teaching and publishing path toward tenure currently do require sacrifice for young professors. Yet an academic career in engineering also offers an extraordinary combination of income and freedom. Students planning their future must be made aware of this.

The battle can be won. Increased economic incentives and a greater national commitment to engineering education are achievable, and many foreign-born experts can be persuaded to remain. Indeed, I am happy to report that John Chen recently decided to stay at Lehigh. Yet others like him may choose differently. We Americans may think of our country as the world's refuge for foreigners, but those of us in academe have discovered that the human pipeline can flow both ways. If it changes its current direction, we are in for some serious problems.

April 8, 1990

Peter W. Likins *is president of Lehigh University and a member of the National Academy of Engineering.*

★ ★ ★

Industrial Cooperation in Japan: It's Not What We Think

George R. Heaton, Jr.

Evidence of Japanese industrial prowess confronts us daily. The Toyotas in our driveways and Sonys in our living rooms soon may be replaced, we hear, by a new generation of Japanese high technology in superconductors and fine ceramics. Scarcely a week goes by without an article comparing our management practices and government policies with those of Japan — usually unfavorably.

It is easy for U.S. observers to portray Japan's system as the harmonious converse of whatever plagues the American economy and to attribute Japan's mastery of technology to the banding together of its government and industry.

This view of Japanese-style cooperation does contain a grain of truth, but it is just false enough to be dangerous. Before rushing to emulate it we should remove our rose-colored glasses and see Japan's cooperative research system for what it really is. I recently spent a year in Japan studying this system and found it more limited than many imagine.

Only a small percentage of Japan's total research and development (R&D) is done cooperatively. The Japan Fair Trade Commission reports that about 55 percent of leading firms join forces in this manner regularly, but 90 percent of the activity involves small-scale undertakings among affiliates. There is also a more visible form of joint R&D, called technology research associations. Since 1961 government and industry efforts have created a network of these associations. Their main purpose has been to bring lagging companies up to speed on technical information known elsewhere.

The impression that the associations produce marketable technology seems widespread in the United States. In fact, their budgets are almost always too small and their goal has been to diffuse existing rather than new knowledge. Also, a spirit of competition — members trying to maximize their own research gains at others' expense — pervades their supposed cooperation. The major value of the associations has been threefold: establishing good government-industry relations, creating information-transfer networks among firms and allowing them to gain a low-cost toehold in new research areas.

Today, many Japanese believe that this system is outmoded. They realize that catch-up is hardly today's industrial agenda and that international linkages are supplanting domestic ties. In consequence, public policy and industrial practices are changing. Greater emphasis is placed on basic science as opposed to applied research, on creating ties between universities and industry and on peer-reviewed awards to individual researchers. Time-honored traditions in the United States, these ideas propel Japan's new technology policy agenda toward one that looks remarkably — and deliberately — like ours.

At the same time, we in the United States move toward policies, such as support for cooperative research, that were the hallmarks of Japan's past. Since 1984, when Congress eliminated antitrust constraints on joint R&D, some 60 private research consortia have formed. The National Science Foundation now funds a network of university-industry engineering research centers, and national labs and private firms work together more easily in such areas as robotics and pharmaceuticals. Sematech, a consortium funded by semiconductor firms and the Defense Department, has set up shop in Texas to work on semiconductor manufacturing.

High expectations accompany these large commitments to cooperative research. Some Sematech proponents, for example, have touted it as the possible salvation of our semiconductor electronics industries.

Japan's experience suggests, to the contrary, that industrial cooperative research serves a much more modest purpose. It can diffuse existing knowledge rapidly. It is efficient for funding basic research, which all firms need but none can own. And it can be a sensible group strategy in the face of a well-understood common threat. These are valuable functions ignored too long in this country. We would undoubtedly benefit from more cooperative research among government, academe and industry, as well as from government funding for industrial research consortia.

But we must also be realistic. Cooperative industrial research is unlikely to produce major technological advances. It will not "save" an industry in trouble because of foreign competition. It should not be supported in the name of national defense. Japan's past and present bespeak an appreciation of these limits. America must take care not to ignore them.

December 27, 1988

George R. Heaton, Jr., *an attorney, teaches at the Massachusetts Institute of Technology and Worcester Polytechnic Institute.*

★ ★ ★

Offering Tools for Soviet Democracy

Paul C. Stern

As Soviet reformers proceed with their great experiment of moving from centralism to pluralism, many Americans who wish them well see only two choices: offering financial relief, such as trade credits, or cheering from the sidelines.

There is another alternative, however, something we can do right now to help the Soviet reformers steer through this difficult transition to a world that will be better for the Soviet people — and for us. We should share our understanding of the nuts and bolts of managing a pluralist, democratic country and a mixed economy.

The Soviet Union faces many problems like our own, including conflict between ethnic groups, bureaucratic inertia and public dissatisfaction with government. But until recently, Soviet social scientists were prevented from examining these problems because Leninist dogma declared they could not exist under socialism. Soviet scholars were prevented from building credible data systems that could tell whether policies were making things better or worse. Instead, they were burdened with fabricated economic statistics, censored census data and dubious public opinion polls.

As a result, Soviet reformers now are setting their course with faulty instruments and ill-trained navigators. If they are to reform their political system and manage their economic problems and ethnic conflicts successfully, they must have better analytic tools and information. They need to train people in internationally accepted research methods and rebuild their systems of social and economic statistics.

Soviet leaders know they need help. They are asking their ethnographers how to get peoples as diverse as the Scandinavian Estonians and the orthodox Muslims of Kirghizia and Azerbaijan to live together in one democracy. They are asking other scholars for advice in developing participatory social and political institutions, coordinating national and local governments, managing factories and privatizing public services. And they are starting to send their students west for training.

U.S. political scientists, economists, sociologists and other social scientists can help, and the Soviets are open to the possibility. For example, shortly after Mikhail Gorbachev initiated his program of *perestroika*, a committee of U.S. social scientists organized by the National Research Council began arranging meetings with social scientists from the Academy of Sciences of the USSR. At the request of the Soviets, the discussions rapidly broadened to encompass public opinion research, ethnic relations and many other social science topics.

Soviet social scientists want to learn how to conduct surveys to get honest answers. One Soviet researcher who recently visited the Center for Puerto Rican Studies at the City University of New York was impressed that American pluralism includes research institutes devoted to assessing the needs of minority groups. Other Soviet scholars have been interested to learn about our methods for objectively evaluating government programs.

We Americans like to complain about pollsters and census takers asking us questions, but their findings provide a knowledge base the Soviets can only wish for. Americans can help the Soviet reform process by advising on ways to build data systems, evaluate policies and interpret the experiences of other countries in managing social problems.

This is not to say that Americans should seek to set Soviet policies, which would be both presumptuous and unwise. Only the Soviets are in a position to understand the needs of their country or to exercise that kind of responsibility. Besides, social science generally does not offer clear answers to policy questions; rather, it helps by giving proponents of different approaches a more factual basis upon which to argue.

Nonetheless, by volunteering to help the Soviets upgrade their statistics and research methods, we can make a real contribution. Soviet leaders can learn which way their experiments are leading, and opposition groups can use the data to keep the bureaucrats honest. More broadly, compiling careful data about the Soviet experience will provide an invaluable resource for others around the world who want to replace authoritarian regimes with democracy.

All of this is not only in the Soviet interest, but in ours. Americans clearly will be better off if the Soviet Union emerges as a more democratic and reality-based society. Less obviously, our own country's systems of politics, economics, conflict resolution and ethnic relations are far from perfect. The bonus from working with the Soviets in this way may be that we end up learning as much as we teach.

March 11, 1990

Paul C. Stern *is the staff officer for a National Research Council project that brings together U.S. and Soviet social scientists.*

★ ★ ★

The Surprising Reality About Hunger

Robert W. Kates

The current famine in Ethiopia has received less attention than Donald Trump's marital problems and new casino.

That's not surprising. Americans contributed generously to relief efforts in Ethiopia several years ago, and they are frustrated to see the problem recur. Recent reports from that country — and from the Sudan, Mozambique and Angola — make hunger appear intractable.

Such discouragement is understandable, but it is misplaced if one considers the larger problem of overcoming hunger worldwide.

The latest news from these parts of Africa is unquestionably grim, and further assistance is needed. Yet, if one looks beyond these examples to the developing world as a whole, the outlook is surprisingly hopeful.

In fact, an international meeting of scientists and other experts on hunger concluded recently that it is possible to reduce worldwide hunger by half during this decade. Doing so would be a remarkable step forward. Those of us at the meeting identified programs and policies that have proven effective in reducing hunger in Asia, Africa and Latin America, as well as in the United States.

Our conclusion was based not on idealism but on a substantial record that has been persistently overshadowed by crises such as the current one in Ethiopia. Famine has been all but eliminated in India and China. Nations such as Zimbabwe show that progress also can be achieved in Africa. For all the bad news we hear, much good has been achieved.

No one is certain exactly how many people suffer from hunger, although the total is at least half a billion persons. Progress in reducing this number was disappointing during the 1980s, but research shows how to achieve much more in the '90s. For instance, programs that provide food subsidies to broad populations through controlled prices or import controls generally have been ineffective. Limited efforts that target the poor with food subsidies, feeding programs and other techniques have been more successful.

In rural areas, providing wage and food income in return for labor on projects to provide needed agricultural and environmental improvements has reduced food poverty immediately while increasing productivity and income. Programs that provide credit to women to start small businesses also have been effective, as have a variety of low-cost techniques that improve agricultural production, provide firewood, limit soil erosion and increase food and income.

These methods work. Similarly, low-cost methods have been developed to eradicate two of the world's major nutritional diseases — deficiencies of iodine and of vitamin A. Many developing countries also have made progress in reducing childhood malnutrition, which afflicts one of every three children. This has been accomplished with immunizations, promotion of breastfeeding, growth monitoring programs and simple techniques to treat diarrhea.

Although most cases of hunger involve specific individuals and families rather than mass starvation, famine remains a serious threat. Yet the toll of famines has declined since World War II, reflecting both an absolute decline and a shift in locale from populous Asia to less-populated Africa. Early-warning systems and emergency food stocks have been put in place to guard against droughts. The famines hardest to alleviate or prevent are those caused by war. However, there is growing interest in the United Nations' becoming more active in protecting civilian food supplies and providing for the safe passage of emergency food relief in such situations.

In other words, a great deal has been learned about how to combat hunger effectively, and many countries have applied this knowledge with at least some success. Famine can be overcome. The stunting of small children and starvation of their mothers can be halted. Even very poor nations can meet their minimal overall nutritional needs.

A systematic assault on hunger inevitably requires additional money and food aid from the rich to the poor and a reversal of monetary flows currently going in the opposite direction. But the sums are relatively small — no more than a 20 percent increase or reallocation of global foreign aid disbursements.

We will not be able to take advantage of this opportunity if we are overwhelmed by despair. News from individual countries may discourage us, but that does not mean the problem as a whole is hopeless. The record indicates otherwise. With renewed social energy and political will, it is possible to cut hunger in half — and then work to eliminate it.

May 20, 1990

Robert W. Kates *is director of the Alan Shawn Feinstein World Hunger Program at Brown University.*

★ ★ ★

Vaccines for the Developing World

Phyllis Freeman

The AIDS epidemic has shown Americans how terrible it is for people to suffer and die for lack of a vaccine. Imagine, then, having vaccines within reach and not developing them.

That is the case today with many diseases that kill, cripple or disfigure people in the developing world. An estimated 14 million children worldwide die each year from malaria, hepatitis, cholera and other diseases. Most of those deaths — and a huge toll in disabilities — could be avoided not only by further distribution of existing vaccines, but also by developing, producing and distributing new ones.

Just as most American parents no longer worry about their children getting polio or whooping cough, parents in developing countries could protect their children from many ailments. This is not just a dream; an enlarged vaccine arsenal is well within our grasp scientifically. The major obstacles are political and economic.

Immunization programs are already operating in most countries, supported by several United Nations agencies. Following the remarkable success of the World Health Organization's (WHO) global campaign that totally eradicated smallpox in the 1970s, the United Nations began the Expanded Program on Immunization, or EPI, in 1974. EPI now protects more than half the world's children against measles, polio, tuberculosis, whooping cough, diphtheria and tetanus — far more than the 5 percent protected in 1974. The WHO and UNICEF are optimistic that 80 percent to 90 percent of all children will be reached with these six vaccines within the next few years.

The six vaccines now used by the EPI were originally produced in — and for — industrial countries. Both public institutes and commercial firms developed the vaccines for their home markets, investing heavily in research because the expected payoff in lives and profits was so great. After recouping their investment costs from sales to industrial countries, many firms were willing to sell the vaccines at or near the cost of production to the United Nations for use in developing countries. On average, EPI pays only 5 cents a dose. The firms have viewed EPI vaccine sales not only as a way to save lives, but also as an opportunity to gain economies of scale, develop new markets and generate goodwill abroad.

This strategy has worked well for the original vaccines, although more children must be reached. However, it is not providing new or improved vaccines against cholera, typhoid, leprosy and other diseases that occur principally in tropical countries. For these, traditional vaccine-makers have little financial incentive to invest, because most of the millions of people who need the vaccines cannot pay for them. Ironically, this dilemma exists at a time when the revolution in molecular biology has provided new tools to create sophisticated new vaccines more quickly.

From a scientific standpoint, vaccines against 19 serious diseases could be ready within a decade if the scientists with the know-how received the support necessary to work on these problems. As reported by the Institute of Medicine of the National Academy of Sciences in 1986, this list includes such miserable diseases as leprosy, childhood diarrheas, pneumococcal pneumonia and meningitis.

The economics are discouraging, but it is no time to declare defeat. The United Nations is exploring two initiatives to overcome these barriers and develop vaccines intended primarily for the Third World. One approach calls for the United Nations to contract with existing public institutes or with commercial firms to do the essential research on specific vaccines. Having paid these development costs, EPI hopes to buy the vaccines later at near the cost of production.

Another initiative, supported by the Pan American Health Organization and the Rockefeller Foundation, aims to develop regional institutes that will work on vaccines for deadly diseases in Latin America. If it succeeds, similar efforts could be launched in Asia and Africa.

These and other alternatives are promising. Yet none is likely to achieve full success without substantial support — both financial and scientific — from the United States and other industrial countries.

It is hard to imagine a more meaningful challenge for the United States to assume as a world leader. By the end of this century we could help slash the incidence of some of the world's most devastating diseases, saving millions of lives. Improved child survival rates would also help ease population pressures; where death rates have fallen in the past, pregnancy rates have followed suit. An opportunity of such magnitude cries out to be exploited.

March 7, 1989

Phyllis Freeman *is associate professor and chairman of the Law Center of the College of Public and Community Service at the University of Massachusetts at Boston.*

★ ★ ★

Easing the Fear of Giving Birth

Julie DaVanzo

A new baby.

Those three words evoke feelings of tenderness in most Americans. In many other countries, however, the prospect of giving birth also evokes another emotion: fear.

In much of Africa, a mother's lifetime risk of dying during pregnancy and childbirth is greater than one in 25. In southern Asia, maternal deaths claim at least one mother in 50. Worldwide, an estimated half-million maternal deaths occur each year, 98 percent of them in developing countries. In our own country, by comparison, fewer than one woman in 10,000 dies from complications of pregnancy or birth.

The problems of developing countries are hard for those of us in the United States to grasp emotionally. We read about famine or a military coup and it seems remote from our own lives. Yet anyone who has ever given birth, or whose wife or sister has, can understand what it means for a woman to die during what should be a joyous occasion.

There is a way to reduce this tragedy significantly, but it requires us to stretch our minds and think in a new way about a controversial topic.

The topic is family planning. When Americans talk about family planning in developing countries, they almost always tie it to the problem of population growth. The discussion is filled with references to China, India and teeming masses, and it has become familiar to everyone regardless of where they stand on the issue. The other time one hears about the subject is when someone ties it to the related, but distinct, issue of abortion.

In fact, many political leaders and health experts in developing countries are most interested in family planning as a means of safeguarding the health of women and children. More than 13 percent of the children in the developing world die before age 5, compared with less than 2 percent of those in countries like ours.

I recently co-chaired a group that studied this problem for the National Research Council, and we found that these local leaders are right. Family planning is essential to the health of mothers and children in the poorest parts of the globe.

Why? First and foremost, because it can reduce the number of times a woman in a remote village or crowded *barrio* exposes herself to the possibility of a ruptured uterus, postpartum hemorrhaging and other complications of pregnancy and delivery. These dangers vastly outweigh any risks of practicing contraception. Family planning enables a young teenager to avoid motherhood until her body matures and spares an older woman the special dangers she faces from pregnancy. It enables other women to increase the interval between births, recovering their strength and reducing the competition among their children for food and care.

In countries in which safe abortion is unavailable, family planning services are especially important for reducing the number of unwanted pregnancies that might otherwise lead to maternal death or injury from dangerous abortion procedures.

Over the past 40 years, many developing countries have achieved significant declines in fertility levels; 400 million couples in these countries now practice contraception. Usage varies from less than 20 percent of married women aged 15-49 in Pakistan and Kenya to more than 60 percent in Colombia and Korea. Overall, however, the incidence of poor health and mortality among mothers and children remains unacceptably high. Those countries with the highest death rates have some of the highest fertility rates. They need expanded family planning services desperately.

Family planning alone will not end the tragedy of so many women and children dying. Also needed are expanded prenatal care, better midwifery, and safer delivery procedures — not to mention a higher standard of living. But family planning is an essential part of the package.

Those of us in the United States can help in this process. Our country generally has supported family planning in developing countries, if sometimes reluctantly. But we also have tended to discuss it in terms of our own concerns — overpopulation and abortion — rather than those of the people involved. For them, family planning is less a matter of demographics or ideology than of health. It's time we recognized this. The half-million maternal deaths that occur each year lie in the gap between our perceptions and their reality.

July 8, 1990

Julie DaVanzo *is senior economist at The Rand Corp. in Santa Monica, Calif.*

★ ★ ★

New Crops for South America's Farmers

Hugh Popenoe

Imagine a delicate stew with the smell and taste of roasted chestnuts, a crunchy and nutritious new snack that looks like popped beans, or a fruit as smooth as ice cream with a taste of pineapples and strawberries.

These foods exist but most Americans have never tasted them. Ironically, they are grown in the same part of the world that now sends abundant supplies of our country's most undesirable agricultural import, namely coca in the form of cocaine.

Peru, Bolivia and Colombia cultivate more than just coca and coffee. A panel of the National Research Council, which I chaired, reported recently that these countries also are home to some of the world's most delectable but little-known foods. Many of these traditional Andean crops are as delicious as they are productive. With technical and marketing support, they might gain popularity among Americans, just as the kiwi fruit has done over the past few years.

This would not solve the drug crisis, which is caused by a complex array of social and economic forces. As an agronomist, I have no special expertise on how to stop the flow of cocaine into the United States. Nor do I suggest that these plants can simply replace coca. Yet, if and when coca production is somehow cut back in South America, farmers there are going to need financially attractive alternatives. Exporting these unsung crops might provide an option for some of the farmers while opening a culinary treasure chest to gourmets worldwide.

Most of the foods were staples of the Incas until the Spanish conquistadors arrived in 1531. The Spaniards overpowered the Incas and forced them to grow such European crops as wheat, barley and carrots. With the exception of a strange plant called the potato, the traditional Inca crops were mostly ignored by the new rulers.

Drawing by Sam Capuano
Copyright, *The Cleveland Plain Dealer*, Ohio

Many of the plants could enjoy profitable niches at supermarkets and health food stores around the world. For example, *arracacha* (pronounced a-ra-CATCH-a) is a carrot-like root whose delicate flavor is a blend of cabbage, celery and roasted chestnuts. With its smooth skin and varied colors, it can be boiled, fried or added to stews. Brazilians have begun discovering arracacha's distinctive flavor; why can't Americans?

Ulluco (oo-YOU-co) is a potato-like tuber whose waxy skin comes in such bright shades of yellow, pink, red, green or even stripes that it looks almost like a plastic decoration. Rich in vitamin C, ulluco has a silky texture and a nutty taste.

Mexican *amaranth* is already gaining popularity in some parts of the United States as a nutritious ingredient in breakfast cereals, granola and other products. Yet most Americans have never heard of amaranth's South American counterpart, even though its popping quality makes it a potential competitor with popcorn and its nutritional value is almost unsurpassed.

New species of peppers are another possibility for gourmets who are on the lookout for exotic spices. The *rocoto*, for example, is fat like a bell pepper but pungent like a hot chile. The *Andean aji* has a subtle, unique flavor that can be blended into sauces.

There also are many Andean fruits. The *cherimoya* is the one that tastes like pineapples or strawberries. The *pepino* is a shiny yellow-and-purple fruit whose flavor resembles a sweet melon. The tangy *tamarillo*, a "tomato" from a tree, can be found in some U.S. markets, but not widely. In New Zealand, it is almost as popular as kiwi fruit.

Consumers in the United States and other countries might open their mouths — and their wallets — to try these and other Andean delicacies. Before that can happen, however, more is needed than just handing out seeds. Scientific development of the crops is required, and an extensive export and marketing system must be established. Farmers need capital and technical support. Most of all, regional problems of drugs, poverty and political instability must be alleviated.

Still, at a time when ethnic restaurants and gourmet stores

are thriving in our country, there should be a way to pique U.S. interest in *arracacha, ulluco* and c*herimoyas*. Rather than just condemning Andean farmers who grow coca, we ought to help them work toward the day when they can turn their magnificent botanical heritage to their own advantage — and ours.

November 5, 1989

Hugh Popenoe *is director of the Center for Tropical Agriculture at the University of Florida, Gainesville.*

★ ★ ★

8

DIFFICULT CHOICES

Life and Death: More Than an Expert Opinion

Ralph Crawshaw

Suppose you were a state health director and you had two proposals before you to spend $10,000. The first proposal would provide a coronary artery bypass operation for one patient; the second would immunize 5,000 children at a cost of two dollars each.

Which would you choose?

What if the choice were between a high school education program for preventing teenage pregnancy and a helicopter ambulance service for sick babies? There is no objective "right" answer in cases like these. You can commission an analysis of each service to determine what it will cost and how many lives it will save, but deciding how to apply the data depends on your values.

Unfortunately, we Americans do a poor job of confronting such choices, much less making decisions about them. We wage fierce debates about weapons programs, taxes and other issues, but when it comes to life-and-death choices involving health care, the silence is deafening.

Perhaps people believe we are able to spare no expense as a society to save lives. Yet choices are inescapable, as we Oregonians learned not long ago when a local boy died from leukemia. The bone marrow transplant he needed was disallowed by state Medicaid officials because the money was earmarked to provide prenatal care to the poor.

Similar choices exist nationwide. Is it right for hospitals to offer luxury rooms and gourmet meals for some patients while 37 million other Americans lack adequate health insurance? Should money from community health programs be diverted to provide expanded medical services for AIDS patients? Are old people entitled to organ transplants?

In a democratic society, value-laden decisions like these should be made by the people or by their elected representatives. Yet, with the notable exception of abortion, Americans generally have had little to say about such issues. They leave the choices to physicians and government officials or, more often, allow them to be resolved by default.

The irony is that Americans have become much more assertive about making life-and-death decisions affecting themselves or their families. Many cancer patients, for example, now insist on being the ones to choose whether they will undergo intensive chemotherapy. Some families of brain-dead patients demand the right to decide whether to keep the person alive with medical technology.

Of course, one of the main reasons Americans are more vocal about these kinds of decisions is that they are affected so personally and directly. Yet it also is true that larger social issues are rarely framed in a way that invites public discussion. Most citizens lack the information or forum to grapple with these questions, much less to help decide them.

An emerging grassroots "health decisions" movement in several states demonstrates that citizens *can* get involved constructively. Citizens' groups in these states have begun discussing as a community many of the tough decisions that burgeoning medical technology has forced upon us. In some cases, they have affected official policy.

For example, Oregon Health Decisions, a citizen-based organization, trained a cadre of 75 homemakers, insurance salesmen, firemen and others in the language of bioethics and health-care decision making. Members of the group then held 300 meetings across the state to talk with other citizens about such issues as death with dignity and equitable access to health care. In response, the state legislature established by law a process that incorporates the health values of citizens in these critical decisions.

One result has been that Oregon has adopted some explicit policies about where it will spend its resources; prenatal care, for example, comes before organ transplants. Some critics have attacked this approach as "health rationing" and called it a dangerous departure from the concept that society should try to provide optimal health care for everyone.

However, I think this new movement is welcome. Our country now spends a far greater percentage of its gross national product for health services than any other nation, yet it has not bought itself a commensurate amount of better health. Many American citizens have severe difficulty obtaining basic health services. Choices *do* exist, even if we prefer to pretend otherwise. We need to confront these decisions squarely.

November 19, 1989

Ralph Crawshaw, *clinical professor of psychiatry at Oregon Health Sciences University, is on the board of directors of Oregon Health Decisions.*

★ ★ ★

The New Diagnostics and the Power of Biologic Information

Dorothy Nelkin and Laurence Tancredi

A company considers enrolling one of its brightest young female employees in an expensive training program. Before making the investment, it asks the woman to undergo biological testing with the latest diagnostic techniques, including genetic screening.

The tests reveal an unexpected problem. The woman carries the gene for Huntington's disease, a degenerative, invariably fatal brain disorder. Although she now seems in

Drawing by Pat Bagley
The Salt Lake Tribune, Utah

good health, she is likely to begin developing symptoms of the illness at about age 40.

Faced with the prospect of paying the huge medical bills, the company decides not only to withdraw the woman from the training program, but to dismiss her entirely.

Situations such as this are becoming possible as new biological tests emerge from the laboratory. Designed to uncover latent problems or predict future diseases, these diagnostic techniques offer potentially valuable clinical applications. Physicians can apply them to identify potential problems and to recommend therapy or preventive action with greater speed and confidence. New brain-imaging methods can be used to diagnose potential behavioral disorders, learning disabilities and psychiatric illnesses. Genetic tests can help identify patients with a predisposition to hereditary diseases and complex disorders suspected of having a genetic component, such as mental illness, Alzheimer's and susceptibility to alcoholism.

A physician might use this information to improve the quality of medical care by warning someone with a genetic predisposition to heart disease to eat certain foods or by cautioning someone prone to certain cancers to receive more frequent checkups. However, the new techniques also may find their way into contexts where they provide unprecedented threats to our traditional concepts of privacy and personal autonomy.

For example, schools, employers, insurers and the courts, concerned about controlling costs or improving efficiency, could use the tests to learn about the health status and behavior of their clients. Such information could limit access to insurance or health-care facilities or exclude high-risk individuals from jobs or training programs.

Within the legal system, new tests can be used to enhance certainty in controversial decisions, such as whether to sentence someone to prison. Legal scholars writing on biological psychiatry predict that courts increasingly will use information from brain scans to evaluate the sanity of criminal defendants and to predict the likelihood of future dangerousness.

Grounded in science, the new diagnostic technologies are compelling. Images on a screen appear precise, and statistical findings processed by computers seem value free. Yet the information produced by most tests is only inferential. Interpretation rests on statistical definitions of "normal" and may assume a cause-and-effect relationship where there are only correlations. The error rate may be high. Moreover, even if tests reliably anticipate who is susceptible to a disease, they cannot predict the age of onset, the severity of expression or the influence of intervening factors.

Despite these limits, biological testing is a growth industry. Biotechnology companies hope that most people will have genetic profiles on record by the year 2000 and that testing will be mandatory in many organizations.

The implications are profound. The refinement of tests already is expanding the number of disease categories and the number of people judged deviant. Just as improved sensitivity in the technologies used to test food products has expanded the number of products identified as carcinogenic, so are improved diagnostics increasing the number of people defined as abnormal.

Predictive testing thus opens possibilities for new forms of stigmatization and discrimination and encourages social policies based on biological criteria. Indeed, some asymptomatic people suspected of having a genetic disease already have been barred from insurance or employment. One can imagine families demanding information about the biological status of relatives, adoption "brokers" probing the genetic history of babies or commercial firms storing genetic profiles and selling the information to insurers or employers.

Considering the rapid development of the new diagnostic technologies, much more discussion is needed about critical questions of privacy, access to information or the potential for abuse. As tests extend the range of what we can predict, we should not assume their moral neutrality. The technology has the potential to serve society well and save many lives. However, its indiscriminate use could define more and more people as unemployable, uneducable and uninsurable — creating, in effect, a biologic underclass.

February 4, 1990

Dorothy Nelkin *from New York University and* **Laurence Tancredi** *from the University of Texas Health Science Center are the authors of* Dangerous Diagnostics: The Social Power of Biological Information.

★ ★ ★

Harvesting Organs from Anencephalic Infants

Alexander Morgan Capron

Every year several hundred children in the United States are born, alive, with anencephaly — that is, without cerebral hemispheres and the top of the skull. Some physicians have asked: why not donate organs from these babies to save the lives of other children, since an anencephalic child will die soon anyway and cannot feel pain?

In response, legislators in several states have proposed laws allowing hearts, kidneys and other organs to be removed from these infants before they meet the usual criteria for death.

This may appear to be a merciful approach to a tragic situation. Yet it is likely to fail and could jeopardize other transplantation programs that have a much greater chance of success. More basically, it threatens the inviolability of human life, producing moral confusion that outweighs any possible benefit.

The idea of relying on anencephalic newborns as an organ source is not new, but it has been seriously pursued only in recent years as transplantation methods for young children have improved.

The first problem with it is simply that the supply of usable organs is likely to be meager. Thanks to prenatal screening and other medical advances, the number of anencephalic births has been declining steadily. This amount would be further reduced because some parents refuse to allow transplantation or because the organs cannot be matched with a compatible recipient. In fact, a realistic annual estimate is that only nine hearts, two livers and no kidneys would be transplanted successfully.

Thus, the number of usable organs from anencephalics would remain far short of the need, even if laws were relaxed to allow transplants to occur before the infants met the standards now used to determine death in other persons. Those standards require the irreversible cessation of functioning in the brainstem as well as in the higher centers of the brain. Yet waiting until an anencephalic's brainstem ceases functioning usually renders the vital organs useless for transplantation because those organs are damaged by the same process — episodic cessation of breathing — that causes the brainstem to fail.

One suggestion to overcome this problem — allowing organ harvesting before death — may be well-intentioned, but is also chilling. It says homicide should be lawful to save

Drawing by Charles Hazard
The Baltimore Sun, Md.

the life of another whose life is not threatened by the person killed, provided the latter is about to die anyway.

But if these principles apply to anencephalics, then they are also valid for other patients. Why not allow relatives to donate organs from patients in the final stages of any terminal illness or from prisoners on death row? Since anencephalics won't supply enough organs, this utilitarian argument could push society to extend the category to children with other terminal diseases to benefit babies awaiting transplants.

Some proponents of using anencephalic infants as organ donors argue that because these children lack consciousness, they are not persons, which eliminates the main ethical problem in killing them. Yet research on both humans and animals suggests that a child with only a brainstem may experience more sensation than physicians traditionally assumed.

Another rationale is that the large majority of anencephalic infants die within a few days anyway. Yet society must scrupulously maintain the distinction between "near death" and death, even if this requires forgoing some organs. Otherwise, the public may suspect that physicians change the rules for transplantation whenever it suits their needs.

Finally, anencephalics who are maintained medically for possible transplantation turn out not to die as rapidly as surgeons require in order to perform the procedure successfully. Indeed, at one transplant center, life support was withdrawn from "possible donors" after one week, but several survived for weeks longer. Also, the criteria for diagnosing when death actually occurs never have been validated for newborns, especially ones with neurological problems of the type anencephalics have.

Given the many problems involved, using anencephalic infants in this fashion is one "breakthrough" society should leave untouched. Parents of anencephalics should be offered the opportunity to donate their child's non-vital organs, such as corneas and heart valves, for transplantation after death, and to cooperate with postmortem investigations aimed at learning more about this tragic affliction. But encouraging them to donate vital organs from their infant will merely intensify their agony or involve them in a decision that they ought not be asked to make: whether one life should be ended to benefit another.

May 6, 1990

Alexander Morgan Capron *is University Professor of Law and Medicine at the University of Southern California.*

★ ★ ★

HIV Screening and the Calculus of Misery

Ronald Bayer

A fierce debate has been taking place within the medical community about whether to test more — or all — pregnant women to see if they are infected with HIV, the virus that causes AIDS. Proponents say such screening would identify women and newborns who need special care while also providing valuable data about the course of the epidemic. Opponents worry about violations of privacy, discrimination and stigmatization.

A critical assumption underlies this debate, however, namely that those who test positive will be provided with the counseling, drugs and other services they need. Otherwise, why would anyone agree to be tested? This issue was discussed at a recent conference on HIV screening held by the Institute of Medicine of the National Academy of Sciences.

Some public health experts, alarmed about the continuing spread of the AIDS virus, argue that screening might have a positive effect by identifying more patients with HIV and increasing the pressure on government hospitals and clinics to provide services. In other words, it would drive the system to respond.

The assumption sounds reasonable — until one considers the plight of homeless families and the mentally ill wandering the streets of our cities.

When homeless people started appearing in large numbers in the 1980s, most compassionate Americans assumed something would be done to help them. In many cases, it wasn't. Similarly, many thousands with mental health problems were released from institutions beginning in the 1960s on the premise that they would be provided with alternative services, such as halfway houses. Instead, many ended up on the streets. Need has not created supply.

There is an established ethical principle in medicine that screening for a disease should not be done without first assuring backup services. It is wrong to subject people to possible anguish and discrimination without offering them the benefits of treatment. Yet, in the face of society's policies towards the homeless and the mentally ill, as well as the current inadequacy of services for AIDS patients, one must question whether such services would be provided to all those who tested positive in an HIV screening program.

The potential benefits of screening are clear. Women who tested positive might benefit from receiving the drug AZT in some cases, and they could be informed that their unborn child has a one-in-three chance of being infected with the AIDS virus. Some might elect to terminate the pregnancy if, in fact, they could obtain an abortion. Others would be alerted to the need to watch their newborn closely for possible medical problems. Yet many — especially those who are poor and from minority groups — would have difficulty obtaining medical services of any kind. Our country's health care system is the most inequitable in the Western world, with 37 million Americans lacking health insurance.

Then there is the question of informed consent. Medical testing for HIV should be done only after the patient provides specific agreement to the test. Fortunately, as the result of aggressive action by white, middle-class gay men, this requirement has come to be accepted by many medical professionals who now deal with the AIDS epidemic.

Yet "informed consent" is easier said than done. Some clinicians who have little experience with HIV and are used to "presumed consent" by patients for diagnostic tests may have difficulty accepting the standard of specific consent, particularly as the epidemic expands into poor patient populations.

All this argues against implementing a widespread, routine HIV screening program for pregnant women. Yet such a conclusion leaves me uneasy. It makes screening appear sinister, which simply is not the case. Its potential value is considerable. Screening could inform many individuals about the need to obtain treatment and avoid infecting others, about the need to consider all reproductive options.

The real villain is the lack of services. Persons considering a test for HIV should be certain about confidentiality.

They must be sure that being diagnosed as seropositive will not lead to discrimination but to adequate medical care. Instead, physicians and patients alike are being forced to make excruciating choices about medicine, justice, life and death. It is a calculus of misery. Widespread screening is no substitute for providing adequate resources.

July 22, 1990

Ronald Bayer *is associate professor at Columbia University School of Public Health and author of* Private Acts, Social Consequences: AIDS and the Politics of Public Health.

★ ★ ★

Laboratory Experiments on Animals Should Continue

Norman Hackerman

Presidential elections and other dramatic news may capture the headlines, but one of the most profound events of our time has been the dramatic increase in life expectancy. Americans now live about 25 years longer than they did a century ago.

That is an essential fact to keep in mind when one considers whether it is ethical to carry out experiments on animals, a controversy that has existed for more than a century. The debate has become more charged over the past decade, with some radical animal-rights activists breaking into laboratories to "liberate" animals.

A number of individuals have argued that progress in computer modeling and other technologies now makes it possible to test drugs, surgical techniques and other medical advances without using animals. The use of animals in experiments, they say, can be reduced dramatically, if not eliminated. On the other side, most scientists warn that animal experimentation continues to be essential to biomedical research.

Who is right? I recently chaired a committee of the National Research Council that tried to answer this question. Our committee, which included experts in biology and generalists such as myself from unrelated fields, spent three years reviewing the evidence.

Our conclusion was that laboratory animals are likely to remain a critical part of human health-care research for the foreseeable future. Researchers have developed a number of experimental techniques that help reduce the need for animals; such efforts should continue to be encouraged. However, there is little chance these alternatives will eliminate the need for research animals anytime soon. Available models simply cannot re-create the complexity of real living systems.

Animals will continue to be needed by researchers seeking ways to cure, treat and prevent such health problems as cancer, AIDS, Alzheimer's disease and diabetes. In the past, animal experimentation resulted in vaccines to prevent polio, smallpox, measles and a host of other diseases. Other Americans owe their lives to kidney transplants, heart-bypass operations, the removal of brain tumors and CAT scans, all of which were tested for reliability on animals before being used with humans.

In other words, science must continue to use animals in experiments if one accepts the moral premise that humans are obliged to each other to try to improve the human condition. Most individuals would reject the contention of animal-rights advocates that it is "speciesism" to use animals in this process. The fact is that we humans are different from other animals. We are the only ones who appear able to make moral judgments and engage in reflective thought. For countless centuries we have used animals for food, fiber, transportation and other purposes. Experimentation is part of this tradition, and it is equally justifiable.

Of course, doing so carries with it a responsibility for stewardship. Chimpanzees, dogs, rats and other animals must be treated humanely. Researchers who do not follow regulations that provide for this should be punished legally and subject to censure by their peers.

The great majority of the animals, however, are treated humanely. Some opponents of animal experimentation allege that abuse and neglect of animals are widespread in research facilities, but our committee could find only limited evidence to support them. Although continual review of the matter is warranted, abuse of research animals is the exception rather than the rule.

Another point of contention has been whether pound animals should be used in research. Twelve states have passed laws that impose prohibitions on doing so, even though fewer than 200,000 dogs and cats are released from pounds and shelters each year for use in scientific research. This is far fewer than the more than 10 million animals destroyed at pounds and shelters annually.

The result of these restrictions, unfortunately, has been to increase the demand for research animals from dealers. This causes more animal deaths overall and raises the costs of research.

Our committee did not examine whether it is ethical to use animals for other purposes, such as to test the safety of cosmetics. Our task was only to review the use of animals in biomedical and behavioral research. We concluded that this kind of research is necessary and of great service to humanity. It should continue.

November 29, 1988

Norman Hackerman *is president emeritus of Rice University.*

★ ★ ★

Integrity and Science

Arthur H. Rubenstein and Rosemary Chalk

A few cases of scientific fraud — in which scientists fabricated research data or plagiarized the work of others — have generated extensive publicity over the past decade. These cases have provoked public concern about the integrity of biomedical research and the ability of the academic research community to prevent future occurrences of misconduct.

Evidence of the extent of scientific misconduct is still lacking, as we learned while examining the problem recently for the Institute of Medicine of the National Academy of Sciences. Our study panel concluded that serious research fraud is most likely a rare phenomenon. Of greater concern are sloppy and careless research practices, which appear to be more common.

This carelessness includes practices such as listing one's name as a co-author of a publication without even reading it or failing to observe accepted research standards in conducting scientific experiments or interpreting findings. A researcher facing increased competition for a grant to continue a medical study, for example, may be more tempted than in the past to exaggerate the significance of preliminary findings.

Such actions, while not fraud, erode the integrity of science and contribute to an over-permissive research environment that fails to discourage more serious forms of misconduct. Given increased funding pressures and the current overemphasis on publication as the main means of achieving status in science, it is time for research institutions to act more forcefully to strengthen the professional standards for scientists.

Traditionally, the research community has guarded against misconduct or sloppy research in two ways. First, scientific papers are usually reviewed before publication by several experts in the field. A problem with this process of peer review, however, is that the reviewers must trust the raw data as presented by the researcher. It may be almost impossible to identify an author who fabricates data convincingly.

The other safeguard is the scientific tradition of repeating experiments to verify another researcher's results. But in an era of complex and expensive research, replication may not occur for most experimental work.

Peer review and replication are still fundamentally sound as a way of assuring integrity in science. But they now need to be supplemented with more formal guidelines and professional standards. Research institutions must establish and enforce these rules on their own and ensure that they are effectively communicated to the next generation of scientists.

For example, research institutions should adopt explicit standards on such matters as how research data should be recorded and retained or who is to be included on the list of authors for research papers. To ease the "publish or perish" pressures that can lead to fraud, universities should join with scientific journals and funding agencies to limit the number of publications they will consider when making decisions on career advancement and research funding. The primary message to researchers should be that quality counts much more than quantity.

To make these changes work, research institutions need not only clarify their expectations about professional conduct, but also provide researchers and students with formal instruction about these rules. They should designate someone who is clearly responsible for promoting high standards of research within the institution and establish an effective procedure for taking action against those who violate them.

The federal government, for its part, needs to assume a more vigorous role in promoting responsible conduct in biomedical research. As the nation's primary funder of health-related research, the National Institutes of Health should take the lead by establishing an office with the responsibility of helping institutions develop the guidelines outlined here.

We oppose another possible federal action that has been discussed, namely random "audits" of investigators' laboratories by government officials. Not only would this expensive approach probably fail to detect fraud, but it could too easily damage the spontaneity and creativity that are essential to a productive research environment; the cure might be worse than the disease.

Adopting these specific rules of conduct will boost everyone's confidence that both honest mistakes and outright cheating will not go unnoticed. Best of all, they will enhance the productivity and reputation of the vast majority of researchers who remain honest and dedicated in their efforts to improve our national health.

March 28, 1989

Arthur H. Rubenstein, *chairman of the Department of Medicine at the University of Chicago, chaired a committee of the Institute of Medicine that studied biomedical research misconduct.* **Rosemary Chalk** *was the committee's study director.*

★ ★ ★

9

THE NEXT GENERATION

Kindergarten Stress

Eugene E. Garcia

In the recent movie "Parenthood," Steve Martin portrays a father who agonizes over the indelible scars his actions may leave on his children. He is obsessed with the thought that a wrong move on his part during his children's early lives will wreak educational and social havoc during their adult years.

The movie rings true because so many American parents share his parental paranoia. From waiting lists at the "right" nursery schools to stores stocked with flash cards for preschoolers, many American parents are putting pressure on their young children to succeed. Parental anxiety once reserved for high school grades and SAT scores is trickling down to children in overalls. Schools, too, are placing greater emphasis on testing and tracking at an early age, labeling children as winners and losers before they even have learned their multiplication tables. Call it "academic trickle-down," which results in kindergarten stress.

Recently, at the request of the National Forum on the Future of Children and Families, I led a group of educators and scholars that examined this growing pressure on youngsters. We found that much of what is happening is misplaced and harmful. Although promoted with the best of intentions, it counters what research by psychologists, education specialists and others has shown to be the most effective techniques of teaching and evaluation.

The growing use of kindergarten screening and "school readiness" assessments provides a good example. Schools in most states now screen children prior to entering kindergarten and first grade. This may seem like a good way of protecting certain children from a situation for which they are not yet ready, but the tests present severe problems. Their validity for identifying those who are developmentally at risk is limited, at best. Recent evidence documents convincingly that between a third and a half of youngsters tested by widely used standardized measures are misclassified. Teacher-made screening tests used by many local school districts are of similarly dubious quality.

Despite this, the use of these tests is growing, with schools emphasizing drills and work sheets to help students score higher. The potential effect of this misuse of tests on how young children learn and develop is chilling.

The very concept of requiring young children to meet specified standards assumes that they should develop in a predictable pattern. Yet research has shown that the early years of human development are highly individualized, with children exhibiting unique learning styles. Tracking them into homogenous groups may make teaching easier, but there is no evidence that doing so is academically beneficial to the children. Recent evidence points in the opposite direction, particularly for children tracked into low levels from which there often is no escape.

Nonetheless, many schools have begun to assign children to new organizational configurations such as "developmental" kindergarten classes and "transitional" first grades. These exotic names do not fool parents, teachers or students; nor do they reduce the inevitable stigma that results. Research shows that making children take special classes or repeat a grade does not improve their performance. Most of them would catch up to their peers in academics and behavior if they were just treated like everyone else.

Flunking kindergarten or being placed in a special class or program may harm a child's self-esteem, motivation and attitudes toward school profoundly, since even first graders know if they have been placed in the slowest reading group. Parents get the message as well. Like Steve Martin in the movie, they take little comfort from assurances that their child will show the right stuff eventually. His or her failure today is perceived as their failure as parents.

This anguish is largely unnecessary. Strategies exist that research has shown to be more effective in teaching reading and math to young people. These alternatives are more expensive and labor-intensive, but they also require less testing, tracking and grade retention.

Perhaps our own collective love and concern for our children is to blame, but the discrepancy between what we know to be appropriate and what we are doing as parents, teachers and policymakers is too wide. We are so concerned with results, accountability and efficiency that we have lost sight of the best way to educate our young people.

January 21, 1990

Eugene E. Garcia, *professor of education at the University of California, Santa Cruz, chaired a panel that studied early school testing for the National Forum on the Future of Children and Families of the National Research Council and the Institute of Medicine.*

* * *

Abe Lincoln's Schoolroom

Philip and Phylis Morrison

Back in Abe Lincoln's rural Indiana school, two things were clear: what the students brought in preparation and what the school's task had to be. The students were from farm families. They came knowing firsthand about birth and death, about the full moon and the way cloth rubbed where the seam was too thick, about how to lever up a heavy rock, how to sharpen a blade, and how milk soured.

There was a lot they didn't know. They had seen few pictures and almost no books; they badly wanted to learn to read. The world outside was pretty much a mystery, like the past from which their language, laws and beliefs came. The school had to help there. It taught what was needed, and it did its best. It taught Abe to read, write and cipher, and a little about the Mississippi, about George Washington the truthful, and about the Bill of Rights.

Children still come to our schools with plenty to know. They bring a wide visual acquaintance with the world near and far, a flood of images, fact and fiction. They see print everywhere, too; signs, posters, even scrawls surround them; magazines and books are commonplace, with all their pictures. But wood-fired ovens and moonlight and mending are scarce. The staff of life is wrapped and sliced, not fragrant in the oven; the production of most everyday necessities has become either invisible or opaque.

Television has made the wide world familiar to the children. Los Angeles streets and Florida beaches and fearful dinosaurs are all to be seen in colorful image. What is deeply missing is an inner sense of the world's real constraints, of the difference between desire and performance. Pushing a button is not like leaning on a crowbar.

The symbols still need teaching; the three R's, the history, the maps, the tales remain urgent. But they lack any foundation beyond word and image. The schools have a big new task that they have not entirely realized: it is to bring in the hands-on world, the real thing that stubbornly resists or wonderfully confirms what one does. What the children need is to grow plants (and see them wilt for lack of water), to complete the cycle by planting the seed they themselves harvest from the plant they grew. They need to build bridges of soda straws that can hold up the weight of many milk cartons. They need to know which connections between bulb and battery produce light, and for how long.

It would be an error to blame schools for our growing lack of contact with the physical world, but an even bigger error not to do something about it. We are all in this bind together; it is the result of a maturing technological world where production is taken farther and farther from the consumer. The pond is soured by smokestacks far away. Unless we change, our understanding too will escape into the distance.

The capacity to judge when things are right, when they work or when they don't work, doesn't apply only to circuits or other matters of science. It also applies to political programs or to buying consumer goods. What we need to know is what evidence can offer, what sort of hopeful answers leaders can seriously give, and what answers are all but empty. It is an understanding that begins with active experience with the natural and the technological world.

So let us teach our children how to read, write and cipher — but let us also help them explore something of how the material world works. They need to sense through hand, eye and mind the limits of what can be done, and how even within stern limits new opportunities can open.

February 11, 1990

Philip Morrison *is a physics professor at the Massachusetts Institute of Technology.* **Phylis Morrison** *is a specialist in elementary science education. This article is adapted from a talk they gave at the National Sciences Resources Center, a joint science education effort of the National Academy of Sciences and the Smithsonian Institution.*

★ ★ ★

What School Volunteers Can Do

Gilbert T. Sewall

Volunteer activity in American education evokes heartfelt, nearly universal enthusiasm. An estimated 1.3 million adults give time to the nation's schools each year, and their contributions are widely hailed. But the realities of educational volunteerism are more complicated than the image.

At a time of budget deficits and widespread concern about education, Americans need to guard against overestimating what volunteers can accomplish. Volunteers are not a substitute for trained teachers and staff. They are a supplemental resource to those who must carry out the central tasks of teaching and helping students learn.

This distinction is important. Well-meaning people outside education often assume that volunteers provide magical added benefits to schools and thereby ensure school improvement. According to this logic, the more volunteers, the better. In recent years, some congressional proposals have sought to expand school volunteer activity by forgiving student loans. However, an infusion of untrained and indebted young labor is not an educational panacea.

A recent review of volunteer programs across the country by a National Research Council committee confirmed that many volunteers are doing outstanding work. Their ranks have grown beyond mothers at bake sales to include an increasing number of college students, business people and older Americans. Although tutoring remains the most familiar activity, school volunteerism includes everything from the operation of special science programs to Latin instruction, dental services and after-school child care. All this activity constitutes a resource of potentially huge dimensions and one that has been barely tapped.

Yet volunteers alone cannot solve the many problems that face our nation's troubled schools. The schools that need the most help, in fact, generally are the ones least likely to have volunteer programs. At schools with large numbers of poor students, there are fewer two-parent families, families have less free time to volunteer, and human resources in the community are generally limited. Volunteers at these schools cannot be expected to shoulder all the tasks normally performed by social welfare agencies.

Many schools that do have volunteer programs, furthermore, could improve these programs. The committee agreed that the best volunteer programs:

- *Match the needs of the volunteers and schools* and provide volunteers with adequate training so they can complement faculty efforts without impeding or confusing educational activity.
- *Provide rewards for students and volunteers alike.* For students, receiving increased attention can itself promote positive feelings about education, improving self-esteem and encouraging performance. Volunteers — who usually find some private satisfaction in their work — must feel their time and effort are appreciated, respected and recognized.
- *Allow teaching staffs to make better use of time.* Volunteers and paid aides should help trained teachers concentrate their energy and skills where they are needed most. Volunteers can conduct study halls, monitor lunchrooms, oversee field trips or patrol playgrounds. Those with appropriate education and teaching ability can assist with remedial instruction or homework.
- *Increase instructional time for students.* Volunteer programs should try to increase the time students spend on lessons, homework and review.
- *Extend services that districts cannot provide.* Volunteers should provide expertise or skills that schools lack. Computer instruction for students and teachers creates a special area of opportunity for the 1990s.
- *Strengthen bonds between schools and communities.* Volunteer programs should recruit a broad spectrum of able individuals who otherwise might have no interchange with public education. In 1972, about 41 percent of U.S. adults had school-age children; today the figure has dropped to an estimated 25 percent. This trend makes schools less bonded to their larger communities, a situation exacerbated by the increased number of mothers who have joined the labor force.

In some school districts I have seen, administrators have become so defensive that they fear public inspection and the meddling of outsiders. They want any volunteers to carry out frustrating tasks without complaint. A few cynical school officials think of volunteerism as a way to reduce community dissatisfaction with mediocre schools. They seek cosmetic programs to draw favorable press coverage and divert attention from structural deficiencies.

This kind of thinking is counterproductive, and it abuses the altruism that makes volunteerism so distinctive. Volunteers are not a panacea, but they can nourish our country's schools with their skills, commitment and enthusiasm. Their contribution is too valuable to exaggerate — or to squander.

April 22, 1990

Gilbert T. Sewall *is president of the Center for Education Studies and editor of* Social Studies Review. *He served on a National Research Council committee that recently studied volunteerism in schools.*

★ ★ ★

The Challenge of Numbers

Bernard L. Madison

"I never could do math. It was my worst subject."

We mathematicians hear these words often. Ignorance of mathematics is considered a badge of normalcy by many Americans. People may grumble about the teenager at the local fast food restaurant who cannot add a sum without a picture-coded cash register, or shudder at having to compute a percentage, but they tend to see such flaws as unremarkable.

Americans are in for a shock. By 1995, eight of the 10 fastest-growing jobs will be based on mathematics, among them scientists, engineers and statisticians. A majority of the 21 million new jobs created by the U.S. economy overall during 1985–2000 will require mathematics skills and a postsecondary education.

As our country faces the task of educating workers for jobs that cannot even be imagined yet, mathematics is more

THE CHALLENGE OF NUMBERS

Drawing by Joe Escourido
The Ledger, Lakeland, Fl.

important than ever before. The world is changing so quickly that people can no longer rely on a static set of facts they learned in school to get them through the rest of their professional careers. Instead, they must have a strong enough command of the fundamentals so they can adapt constantly. Clerks must learn to become keypunch operators and then systems analysts. Those who cannot learn new skills risk unemployment.

No competency is more fundamental in this emerging world than numerical reasoning, problem-solving and other basic mathematical skills. Informed citizens must be able to cope with economic indicators, census data, environmental risk factors and weather probabilities. College curricula require mathematics and statistics courses. High technology has invaded the workplace.

How well is the United States positioned to meet this increased need for mathematics education? Not very. An expert committee of the National Research Council that studied the situation concluded earlier this month that the nation's supply of mathematically skilled teachers, scientists, engineers and others is actually *shrinking*.

The number of bachelor's degrees awarded in mathematics in the United States was lower in 1986 than in 1966, which helps explain why so many school systems across the country cannot find qualified mathematics teachers. More than half of the new doctorates in mathematics are now awarded to foreign citizens, up from only about one-fifth in the early seventies. By the end of this decade, the number of new doctoral graduates taking academic positions will be insufficient to replace retirees. Even this dismal state of affairs assumes a continued heavy reliance on foreign citizens and makes no allowance for a projected increase in demand for mathematicians from private companies and others.

One reason for the shortage of mathematically trained Americans is that the college-age population as a whole is declining. However, the equation is more complicated than that. Notably, despite some encouraging trends, women, blacks and Hispanics still participate in mathematics-based classes and occupations at rates far below their numbers. Since two-thirds of new workers will come from these groups, the supply of mathematically educated workers is diminishing as the demand grows. That is, unless things change soon.

Mathematics teaching in our country has been hampered by too few resources, lack of a national imperative, a highly decentralized system, unimaginative courses and curricula, and no clear understanding of what is important. At the collegiate level, mathematical sciences have passed through three roller coaster decades since the Soviet launch of the Sputnik satellite in 1957. During the '60s, the study of mathematics and science was equated with supporting democracy, and expanded accordingly. In the '70s, in the face of increasing college enrollments and an emphasis on social problems, the job market for many science and math graduates declined. This past decade, concern about economic competitiveness spurred a partial recovery.

Various organizations are now working on better teaching methods and other promising ideas for the '90s. But the larger challenge exists outside the classroom in our country's offices, factories, shopping malls and homes, where many Americans continue to regard ignorance of mathematics as routine. Such ignorance must be recognized, instead, as a sentence of obsolescence. Everyone, regardless of race or sex, can learn mathematics. Everyone *should* learn mathematics. Until more Americans get the message, our national well-being is very much at risk.

April 15, 1990

Bernard L. Madison *is dean of the Fulbright College of Arts and Sciences and former chair of the mathematics department at the University of Arkansas. He was the study director of a recent National Research Council committee that examined the supply of mathematically trained Americans.*

★ ★ ★

On Darwin, Bibles and Classrooms

Francisco J. Ayala

The Supreme Court recently declared unconstitutional a Louisiana law mandating that creationism be taught in the public schools whenever evolution is taught. Its decision preserved the separation of Church and State, as mandated by the First Amendment to the Constitution.

Did the decision also strike a blow against the cause of religion and religious education? Some critics have said so, denouncing the 7-2 ruling. They're wrong.

Creation science is a concept developed by Christian Fundamentalists that holds that the universe was created suddenly from nothing; that man and the different kinds of plants and animals were created separately; that there was a worldwide flood; and that the earth and living things are of recent origin, rather than millions of years old.

These are notions taken literally from the Bible's book of Genesis, not conclusions reached independently by scientists.

The Court was right to dismiss as a "sham" the pretense that they are scientific. Creation is a religious concept, and "creation science" is not science at all, but an oxymoron, a contradiction in terms.

When scientists talk about the "theory" of evolution, they use the word differently than in ordinary language. In everyday English, a theory is an imperfect fact, as in "I have a theory about why Gorbachev favors *glasnost*." In science, however, a theory is based on a body of knowledge.

According to the theory of evolution, organisms are related by common descent. There is a multiplicity of species because organisms change from generation to generation, and different lineages change in different ways. Species that share a recent ancestor are therefore more similar than those with more remote ancestors. Thus, humans and chimpanzees are, in their structure and genetic makeup, more similar to each other than they are to baboons or to horses.

Scientists agree that the evolutionary origin of all animals and plants is a scientific conclusion beyond reasonable doubt. They place it beside such established concepts as the roundness of the earth, the motions of the planets and the molecular composition of matter. That evolution has occurred, in other words, is a fact.

Not everything in the theory of evolution is equally certain. Many aspects remain subject for research, discussion and discovery. But uncertainty about these aspects does not cast doubt on the fact of evolution. Similarly, we do not know all the details about the configuration of the Rocky Mountains and how they came about, but that is no reason to doubt the Rockies exist.

The theory of evolution needs to be taught in the schools because nothing in biology makes sense without it. Modern biology has broken the genetic code, opened up the fast-moving field of biotechnology and provided the knowledge for improved health care. Students need to be properly trained in science in order to improve their chances for gainful employment and enjoy a meaningful life in a technological world.

Doesn't this pose a threat to religion or to Christianity? Many religious authorities do not think so. Catholic, Episcopal and other Protestant bishops joined Jewish organizations, educators, scientists and civil libertarians as plaintiffs against the creationism law. They saw a threat to their beliefs in statutes that make a mockery of both religion and science with the pretense that the words of the Bible are scientific propositions.

Pope John Paul II told the Pontifical Academy of Sciences in 1981: "The Bible speaks to us of the origins of the universe and its makeup, not in order to provide us with a scientific treatise, but in order to state the correct relationship of man with God and the universe. Sacred Scripture wishes simply to declare that the world was created by God."

The Pope says that it is a blunder to mistake the Bible for an elementary book of astronomy, geology and biology. Instead, it is possible to believe that the world has been created by God while also accepting that the planets, the mountains, the plants and the animals came about by natural processes. I can believe that I am God's creature without denying that I developed from a single cell in my mother's womb. There is no need for warfare between religion and science.

In short, those who seek a literal interpretation of every word in the Bible do not have exclusive rights on Christianity. Neither do they represent the views of most Christian churches.

Fundamentalists are entitled to their beliefs, but the Court was right to insist that the teaching of evolution belongs in the classroom while creationism does not. Evolution is science while "creation science" is simply a subterfuge to introduce Biblical teachings. As President Ulysses Grant said in 1875: "Leave the matter of religion to the family altar, the Church and the private school, supported entirely by private contributions. Keep the church and the state forever separate."

July 5, 1987

Francisco J. Ayala *is professor of genetics at the University of California, Davis, and a member of the governing council of the National Academy of Sciences.*

★ ★ ★

The Long Haul to a Doctorate

Susan Coyle

Universities are opening for the fall term, and young people who received their bachelor's degrees this past spring are coming face to face with their decision to either enter graduate school or go to work right away.

For many bright students, the choice was easy. They decided to spend two years in business school or three years in law school, an investment likely to yield a healthy starting salary and desirable career track.

But for other top graduates, those interested in science and engineering fields, the decision was more agonizing. For them, getting a doctorate takes about seven years, followed by up to three years in a postdoctoral appointment.

Little wonder that many of these graduates are saying, "Thanks, but no thanks." There is a growing decline of Americans pursuing doctorates in scientific and engineering fields, and one reason is that it takes longer and longer to earn a Ph.D.

This is a dilemma not only for the students themselves but for any American who wants new medicines, better transportation, a cleaner environment or new consumer products. A steady supply of doctoral recipients is essential to teach, do research and create the knowledge that private industry uses to develop new products and services. Although graduate enrollments are rising in the sciences and engineering, most of the increase is now due to foreign students — and many of them return to their homeland.

In 1967, it took about five years to earn a doctorate in technical fields. Now it takes two years longer. Since many students take time off during their studies, the mean total time between receiving a bachelor's and doctoral degrees actually is 10 years. During the past two decades, this "total time to the doctorate" has increased by as little as four months in economics to nearly three years in the health sciences, with increases of at least two years in mathematics, psychology and the social sciences.

For students, this means more debt, less income and perhaps postponing the start of a family. Although most scientists do love their work, few are so single-minded as not to consider other career options. If they choose to become lawyers, investment bankers or something else, their skills and insight probably are lost forever to science.

We also are missing the opportunity to diversify the scientific work force by widening our country's traditional pool of technical talent — white male doctoral students. As the number of these students declines in physics, chemistry, earth sciences, mathematics and engineering, more women and minorities are acutely needed to fill the ranks. But their talents, too, will be lost as many of them size up the current situation and head elsewhere.

Taxpayers also suffer. Graduate students pay only about 12 percent of the approximate $25,000 annual cost of their education. The rest generally comes from federal research grants, the budgets of state universities and other public sources. Adding a couple of years to the time required of 13,000 American students adds up to "real money" that otherwise might be spent on financial aid for minority students, new research facilities or other pressing needs.

Some have suggested that this disturbing trend is the result of the additional time needed to cover the explosion in scientific knowledge. After all, there is much more to learn than there was in Thomas Edison's era. Yet this justification fails to explain why students in the same field take such varying lengths of time to complete their degrees. Those with fellowships or research assistantships usually complete their degrees more quickly than others. In fact, students paying their own way often need *five or six years* more to complete their doctorates. In other words, the problem is not scientific complexity so much as financial inadequacy.

The problem goes beyond money to include market forces, university policies, student readiness and other factors. Whatever the reasons, the road to the doctorate has become too prolonged, both for students and for society generally. If our nation wants continued technical advances, it must make it easier for its sons and daughters to get the advanced training they need.

September 9, 1990

Susan Coyle, *a project officer with the National Research Council, co-authored a study with Howard Tuckman and Yupin Bae on the increased time needed to complete doctorates in science and engineering.*

★ ★ ★

The 'Mommy Track' in Science

Paula Rayman

The well-publicized problems of the "Mommy Track" exist not only in the corporate board room, but also in the laboratory. Science, that most rational of pursuits, is making it irrationally difficult for women to succeed.

The science and technology (S&T) community in our country desperately needs to attract more women to help develop medicines, invent products, clean up the environment and improve our industrial potential in the years ahead. Our S&T ranks are now populated primarily with white and foreign-born males.

Most universities and research centers do not actively discourage women from entering technical professions; on the contrary, many of them have admirable recruiting programs. Yet, despite these efforts, the S&T community as a whole continues to make such stringent demands on the time and energy of young researchers that it is extraordinarily difficult for them to raise families while pursuing their careers.

Some Americans may think scientists spend their time

sitting under trees watching apples fall, but the reality is often 60-hour work weeks and all-night experiments. It is a tradition — and an unofficial test of seriousness — that harkens back to a time when husbands worked in the laboratory with wives who raised the children and served as unrecognized, unrewarded lab assistants.

The demands of keeping up a career in science are intensive. The sociologist Robert Merton described more than a decade ago the fierce competition among scientists throughout history to be recognized as "the discoverer." The urgency of original discovery creates a work environment that encourages one-upmanship rather than nurturance, a construction of life built around the devotion to the pillar of work.

Several years ago a Macy Foundation study found that most young scientists of both sexes believe that domestic life is incompatible with productive science. A majority of the male scientists agreed with the view that motherhood marks the end of productive careers for their female colleagues, that "a scientific vocation, like religion, is a calling demanding superhuman (i.e., superman) dedication."

All of the last three women to win a Nobel Prize in the sciences were childless. One of them, Rita Levi-Montalcini, who shared the prize in medicine, wrote in her autobiography that she deliberately chose to forego marriage and children for the sake of her career. Three scientists are too few to constitute a meaningful sample, of course, but their experiences are instructive.

As I discovered recently when carrying out a study of 20 dual science-career families, the women scientists reported a "no-win" choice between family commitment and a "serious" science career. The women, from early 30s to mid-60s in age, held degrees in fields such as medicine, chemistry, mathematics and astrophysics. Like their scientist husbands, most of them reported working at least three evenings a week and often on weekends. However, their work histories and those of their husbands revealed striking differences in job conditions, financial rewards and social support:

- Most of the women reported being actively discouraged

from entering science by a parent, teacher or peer. All of the men said they were encouraged to become scientists.

- Women who had the same occupational ranking as male colleagues reported earning fewer prestigious awards and research appointments. Nationally, women scientists earn 5 percent to 18 percent less income than men scientists of an equivalent rank.
- All but one of the women scientists surveyed reported multiple experiences with discrimination, the most common being sex discrimination, sexual harassment and age discrimination. No male scientists reported such experiences.
- Most of the women scientists reported moving geographically to help their husbands' careers, often more than once. The men reported moving to help their wives' careers far less often.
- Women scientists generally were expected to take primary responsibility for their children, either providing care themselves or "making the arrangements." None of the husbands reported taking primary care of family needs.

Science careers should not be so difficult for women. If the United States fails to attract more women to the sciences, especially in the high-growth fields of math and computers, its economy and society will lose a vital source of innovation.

The S&T community cannot solve this continuing problem simply through recruitment efforts, as worthwhile as these are. It also must confront the hurdles women face in their work lives. The community must develop its own new scientific equation, one that provides a better balance between the challenging work of home and the laboratory.

December 10, 1989

Paula Rayman *is research program director at the Stone Center of Wellesley College in Massachusetts.*

★ ★ ★

Dr. King and Blacks in Science

Willie Pearson, Jr.

Martin Luther King, Jr., is being remembered this weekend for his historic contributions to civil rights. Another of his accomplishments, however, was that he was a black American who obtained a doctoral degree.

That achievement stands in sharp contrast to the situation among black students today, particularly in the scientific and technical fields likely to shape the future. In 1988, for example, of all the doctorates awarded in mathematics and computer science in our country, just *one* of each was awarded to a black U.S. citizen. That is a discouraging commentary on the challenges still facing our nation more than two decades after Dr. King's assassination.

Black American scientists have a proud tradition that ranges from Benjamin Banneker and George Washington Carver to more recent examples like Ronald McNair, the physicist who died in the *Challenger* explosion. Yet the number of black U.S. citizens earning doctorates of any kind has declined since 1978, and the trend is especially pronounced in the natural sciences and engineering. In 1988, black U.S. citizens earned only 95 doctorates in these fields, or 1.1 percent of the total. By comparison, blacks comprise about 12 percent of the U.S. population.

The irony is that our nation's need for this neglected source of brainpower has never been greater to help solve problems in industry, space, the environment, agriculture and other fields.

Our country faces serious shortages of U.S. citizens with doctorates in natural science and engineering disciplines. A growing percentage of these degrees are now awarded to foreign-born students. Between 1962 and 1987, the National Science Foundation reported, the share of doctorates in chemistry, physics and other physical sciences awarded by U.S. universities to American citizens declined from 85 percent to 61 percent. In engineering, U.S. citizens now earn only four of

Drawing by Jill Shargaa
The Orlando Sentinel, Fla.

every ten — or fewer than half — of the engineering doctorates at our universities.

Among those Americans who do receive these doctorates, the great majority are white and male. Yet white males comprise a diminishing share of the total college population, and a decreasing percentage of them express an interest in science and technology.

Blacks and other racial and ethnic minorities, together with women, must fill this gap. By 2010, minorities will comprise nearly 40 percent of Americans under the age of 18. The need is unprecedented to put aside prejudices and draw upon this pool of talent.

Failure to do so will leave the nation little choice but to increase its dependence on foreign-born students. Of course, many foreigners do remain in the United States and make valuable contributions, as Albert Einstein and others have done in the past. Yet this supply source could evaporate quickly as a result of political developments or changing employment opportunities abroad.

Helping more black Americans obtain doctoral degrees also would promote a more secure balance of occupations within the black community itself. As the National Research Council reported in a recent study, a disproportionate number of black Americans currently work in jobs that are threatened by automation or are likely to decline in the future.

What can be done to attract more black students to careers in science and engineering? The answer begins in the earliest grades, with improved pre-school and primary education. As they get older, minority students need encouragement to enroll in advanced math and science courses rather than stopping after introductory algebra or biology. More funding would help historically black colleges and universities, which have been more successful in training blacks for scientific and technical careers. Blacks at other institutions also need increased financial aid, as well as protection from possible racism.

Recent statistics do offer cause for hope. In a survey taken in the fall of 1988, blacks accounted for 11.5 percent of the freshmen at U.S. four-year colleges and universities who were planning to pursue majors in the physical sciences. Yet it remains to be seen how many of these students will fulfill their dream or go on to graduate study. One of the best ways to honor the birthday of another dreamer, Martin Luther King Jr., is to act much more decisively as a nation to help more black students join him as holders of a doctoral degree.

January 14, 1990

Willie Pearson, Jr., *is professor of sociology at Wake Forest University and co-editor of the recently published* Blacks, Science and American Education.

★ ★ ★

The Civilized Engineer

Samuel C. Florman

When I was 21 years old, I found myself on a faraway island in the Pacific Ocean with the U.S. Navy Civil Engineer Corps. In the evenings we sat around drinking beer, playing cards and talking. We were engineers, and mostly we talked about engineering — also about baseball and girls.

The only officer who was not an engineer was the chaplain. One night he joined us at cards, and between hands the conversation was carried on in the usual way. Several times the chaplain tried to introduce a new topic with some intellectual substance: What did we think about the morality of dropping those atomic bombs on Japan? What were our thoughts on religion, art and literature?

Instead of responding, we invariably went back to engineering, baseball and girls. Finally the chaplain slammed down his cards, looked upward and said in a loud voice, "Dear Lord, I know that I am unworthy. I confess that I have sinned, but why did you have to abandon me on this island with nobody for company but these boring, boring engineers?"

I'm sorry to say that the chaplain's view of the engineering profession is widely shared. Many problems in our country, from ugly factories to environmental pollution, are blamed on engineers. Although such difficulties are actually the responsibility of the entire community, it is true that our country's engineers *could* play a more active role in creating technology that is more noble — aesthetically, environmentally and even morally — than what we have now.

Our country also would benefit if its engineers were more active in helping to solve social problems. For the most part, the United States is now run by lawyers and business people. Yet many issues, from national defense to the trade deficit, have a large technological component.

There are, to be sure, a handful of engineers, such as Chrysler

Chairman Lee Iacocca and White House Chief of Staff John Sununu, who are influential in contemporary society. Yet far too many engineers in our country are outside the intellectual, social and cultural mainstream, unable to "speak the same language" as their fellow citizens. That is bad for them; more important, it is bad for society.

It was not always so. Engineers were once among the giants, the leaders of our society. Consider the Roeblings, father and son, who more than a century ago designed and built the Brooklyn Bridge. John Roebling studied philosophy under the great Hegel and wrote essays reflecting profound spiritual concerns. His eldest son, Washington, was a masterful writer of English prose, fluent in French and German, and an accomplished violinist. Although the Roeblings were outstanding individuals, they were not unique.

Many things happened to cause American engineers to become more narrow in their interests, but the most important was the decision a century ago by educators to make engineering an undergraduate field of study, consequently pushing aside liberal arts studies in the curriculum. Before long, most faculty and students came to view the few non-engineering courses as a bothersome waste of time. Certainly, engineering itself is a wonderful manifestation of the human spirit and a vital part of the academic enterprise. But excluding the humanities in this way diminished the benefits that engineers were capable of bestowing upon society.

Happily, there have been some recent signs of enlightened change in engineering education. The Thayer School at Dartmouth College remains the only American school that requires its students to obtain a bachelor of arts en route to a bachelor of engineering, but efforts are under way at many American engineering schools to enrich the liberal arts element of the engineering curriculum.

Realistically, only a select few American engineering students will, in the near future, make their way through these enriched educational programs or attain new leadership roles in society. But maybe a select few is just what we need, at least for a start. In my imagination, I see a day when another chaplain shares an island with a new group of engineers. But these will be Renaissance engineers, and the chaplain, thinking about being in their company, will be thanking the good Lord instead of complaining to Him.

February 25, 1990

Samuel C. Florman *is vice president of Kreisler Borg Florman Construction Co. in Scarsdale, N.Y., and the author of several books on engineering. This article is adapted from a longer version that appeared in* Issues in Science and Technology.

★ ★ ★

Index

A

Abortion
　availability of safe, 173
　contraceptive use and, 147, 148
　for HIV-infected women, 191
Accidents
　prevention of, 94–96
　rate among elderly, 42, 43
Acid rain, 139
Acquired immunodeficiency syndrome (AIDS). See AIDS
Addiction. See also Substance abuse
　to chewing tobacco, 85
　treatment programs for drug, 117–119
Adolescents. See Teenagers
Africa
　childbirth in, 172
　hunger in, 4–5, 167–169
　vaccines for, 4–5, 171
Age discrimination, 218
Agriculture
　alternative systems of, 67–69
　Andean, 175, 177–178
　effect of global warming and climatic change on, 138
　and effect of irrigation, 67, 72–74
　impact of biological advances on, 135
　pesticide use in, 69–73
　use of ocean resources for, 137
Ahearne, John, 14
AIDS (acquired immunodeficiency syndrome)
　behavioral changes to prevent, 96–98
　drug abuse treatment and, 117
　drug experiments for, 99–101
　HIV screening and, 190–192
Air traffic, 30–33
Airbag technology, 95
Aircraft design, 32
Airport capacity, 30–33
Alberts, Bruce M., 144
Alcohol treatment programs, 86–88
Alcoholics Anonymous (AA), 87
Amaranth, 177
Andean aji, 177
Andean crops, 175–178
Anencephalic infants, 186–189
Animals, laboratory experiments on, 192–194
Antarctic Treaty, 62
Antarctica, 60–62
Archimedes, 3
Arracacha, 177

Arsenic, 73
Artificial intelligence, 141–143, 158
Astronomy
  advances in, 134
  impact of light pollution on, 150
Athletes, as role models, 83–85
Audits, random, to detect scientific fraud, 196
Automobiles
  energy-efficient, 139
  safety standards for, 95
  and transportation improvements, 29
Ayala, Francisco J., 211
AZT (azidothymidine), 100, 191. See also AIDS (acquired immunodeficiency syndrome)

B

Banneker, Benjamin, 219
Baseball players, 83–85
Bassuk, Ellen L., 110
Bayer, Ronald, 190
Beach erosion, 136
Becker, Marshall, 96
Benbrook, Charles M., 69
Bible, 211, 213
Biological diversity, 74–77
Biological testing, 183–186
Biology
  of forests, 128–129
  revolution in study of, 133–135
  and theory of evolution, 212
Black Americans. See also Minorities
  in mathematics-related areas, 210
  in science, 219–221
  status of, 107–109
Bookman, Charles A., 136
Boron, 73
Bounded rationality, 123
Brain-imaging technology, 184
Breeding, 63
Bulger, Roger J., 91
Bush, George, 123, 130, 137
Businesses, role in exposing children to science, 5, 6

C

Cancer
  relationship between diet and, 82
  use of chewing tobacco and, 84
Capron, Alexander Morgan, 186
Carbon dioxide
  forests and, 127–128
  increases in air of, 49, 55–56, 139
Carver, George Washington, 219
Cesarean deliveries, 91–93
Chalk, Rosemary, 195
Chen, John, 160, 162
Chen, Kan, 27
Cherimoya, 177
Chewing tobacco, 83–85
Children. See also Infants; Teenagers
  educationally related stress in, 201–203
  effect of child care on, 115–117
  homelessness among, 110–112
  immunization programs for, 170–172
  latchkey, 116
  malnutrition in, 168, 169
  poverty among Black, 107
China
  population growth in, 173
  reliance on coal, 57
Chlorofluorocarbons (CFCs), 55, 61
Cholesterol, 82
Chromosomes, human, 144–146
Cigarettes, chewing tobacco and, 84
Climate changes. See Greenhouse effect; Sea level changes
Coastal deterioration, 49, 50

Coca crop, 175, 178
Cocaine, 175
College education
  foreign, 159
  incentives to pursue advanced technical degrees needed during, 161–162, 221
  See also Doctoral degrees
  science courses for non-science majors, 12, 13
Commodity program in agriculture, 68
Communication
  risk, 14–16
  system and terrorism, 33–35
  use of ocean for, 137
Comparable worth, 112–114
Computer applications
  artificial intelligence and, 141–143
  among elderly, 43
  eliminating animals in research through, 192
Computer industry, U.S., 24–27
Computer software
  for digital switching, 34–35
  international development of, 26
Congress, U.S.
  and budget decisions for science and technology, 152, 153
  elimination of antitrust constraints on joint R&D by, 164
  need for scientists in, 18
  pesticide use issues and, 70, 71
  and radioactive waste disposal, 66
Conservation. See Environmental protection
Construction industry, home, 39–41
Consumer Product Safety Commission, 120
Contraception
  in developing countries, 173–174
  options in, 146–149
Cooper, Theodore, 99
Cooperative research, 164. See also Research
Copernicus, 141, 143
Coronary artery disease, 81–82
Cosby, Bill, 3
Court cases, statistical analyses used in, 120–122
Coyle, Susan, 214
Crawford, David L., 149
Crawshaw, Ralph, 181
Creationism, 211, 213
Criminals
  drug treatment for, 118, 119
  genetic testing of suspected, 185
Czaja, Sara J., 41

D

Dartmouth College, 223
Darwin, Charles, 141, 143
Data systems, Soviet, 166
DaVanzo, Julie, 172
Dean, Robert G., 48
Deep ecology, 74–77
Defense
  use of oceans for, 137
  weapons production for, 37–38
Deforestation, 49, 55, 127
Diagnostic technology, 183–186
Diet, chronic illnesses and, 81–82
Digital switching technology, 34
Disasters
  natural, 44–46
  technological, 14, 15
Discrimination
  age, 218
  and comparable worth issue, 113
  racial, 108, 109
  sex, 218. See also Women
Disease. See also Health
  genetic mapping and, 144–146
  nutritional, 168

relationship between diet and, 81–83
scientific research on, 4–5
screening for, 190–192
sexually transmitted, 147–148. See also AIDS
in trees, 129
tropical, 171
vaccines to combat, 170–172
DNA. See also Recombinant DNA technology
explanation of, 144–145
sequencing, 145–146
structure of, 134
Doctoral degrees
for Blacks, 219
difficulties in obtaining, 214–216
Downs, Hugh, 60
Drainage water, 67, 72–74
Driver education, 95
Drought
global warming and, 55
hunger and, 169
Drug experiments, 99–101. See Substance abuse
Dynamic random access memory (DRAM) chips, 26

E

Earthquakes, 44–46
Ecology, deep, 74–77. See also Environmental protection
Economic forecasts, 122, 124
Economy, impact of biological advances on, 134–135
Ecosystems, 64, 129
Edison, Thomas, 3
Education. See also College education; School; Universities
artificial intelligence used in, 142–143
elementary, and science, 5
and kindergarten stress, 201–203
mathematics, 208–211
preparing and attracting American-born students to study science and mathematics, 161–162, 210
role in modern world of, 203–205
Educational testing, 201–203
Ehrenkrantz, Ezra, 39
Einstein, Albert
advances in physics and, 134
as example of foreign-born scientist, 221
Elderly people
environmental design for, 41–43
hip fractures in, 102
Electric automobiles, 139
Electricity, 140
Electronic fetal monitoring, 92, 93, 102
Elementary school. See also School
exposure to science in, 5
testing and stress in, 201–203
Embryo transfers, 88, 91
Employees, environmental needs for elderly, 42–43
Employment
of Black Americans, 109
civil service, 113
comparable worth issues in, 112–114
parental leave policies and, 116
and use of diagnostic technology, 183–186
Energy Department, U.S., 139, 140
Energy policy
controversial nature of, 56–57
environmental issues and, 138–140
Energy production, use of seabed for, 137

Engineering research centers, 164
Engineering societies, 159
Engineering students
  in foreign universities, 159
  humanities courses for, 223
  lack of U.S., 161
  studying for doctorate, 214–216
Engineers
  broadening role for, 222–224
  foreign-born, 160–162, 219–220
  media's distorted image of, 6–8
  need for awareness of research and achievements by foreign, 158–160
Environmental design, for elderly, 41–43
Environmental protection. *See also* Greenhouse effect
  and alternative systems of agriculture, 67–69, 73–74
  and Antarctica, 60–62
  deep ecology movement of, 74–77
  disposal of radioactive waste and, 65–67
  energy usage and, 138–140
  forests and, 127–128
  genetic engineering and, 62–64
  human needs and, 75–77
  hunger and, 168
Environmental Protection Agency, 70–72
Equal-opportunity legislation, 113
Ervin, Robert D., 27
Ethics. *See also* Medical ethics
  of HIV testing, 190–192
  involving experiments on animals, 192–194
  scientific research and, 195–197
Ethiopia, 167, 168
Evolution, 211–213
Expanded Program on Immunization, 170, 171
Experimental drugs, for AIDS, 99–101

Expert witnesses. *See* Witnesses
Exploration
  seabed, 136–138
  space, 130–133
Extinction
  effect on biological research of, 135
  guarding sea turtles against, 57–59
  of species in tropical rain forests, 75
Extinction spasm, 75

F

Family planning, 173–174. *See also* Contraception
Famine, 167–169
Farming. *See* Agriculture
Federal policies
  agricultural commodity program, 68–69
  and responsible research, 196
Fedoroff, Nina, 62
Fertility
  in developing countries, 174
  treating problems in, 88–91
Fertilizer residues, 73
Fetal research, 90
Fiber-optics technology, 34, 137
Fienberg, Stephen E., 120
Film industry, image of scientists, engineers, and technology given by, 6–8
Fires
  preparation for disaster from, 44–46
  vulnerability of phone system in, 33–34
Fishing industry, 136
Florman, Samuel C., 222
Food
  Andean crops for, 175–178
  eradication of shortages of, 168–169

and health, 81–83
  residues of pesticides in, 67, 71
Food and Drug Administration
  evaluation of contraceptives by, 148
  risk standards of, 71
Foreign-born individuals. *See also* Minorities
  dependence on, 221
  in engineering, 161, 219–220
  in science and technology community, 216, 219
  in U.S. universities, 159
Forests
  biology of, 128–129
  destruction of, 75–77
  as environmental buffers, 127–128
  research needs concerning, 129–130
Fossil fuels
  need to shift from use of, 56, 138
  results of burning, 49, 55
Fractures, hip, 102
Fraud, scientific, 195–197
Freeman, Phyllis, 170
Fuller, Samuel H., 24

G

Galileo Galilei, 3
Gallagher, Susan S., 94
Garcia, Eugene E., 201
Gas drilling, ocean, 136
Genetic factors
  diagnostic testing and, 184–186
  health and, 82
Genetic mapping, 144–146
Genetically engineered microorganisms, 62–64
Gerstein, Dean R., 117
Global warming. *See* Greenhouse effect
Goldman, Steven L., 6
Gorbachev, Mikhail, 130

Gordon, John C., 127
Government officials, risk communication by, 14–16
Grant, Ulysses S., 213
Greene, John C., 83
Greenhouse effect
  energy use contributing to, 138–140
  federal studies regarding, 152–153
  role of forests in, 127–128
  and science literacy, 13
  and sea level changes, 49
  uncertainty surrounding, 55–57
Gynecologists, 91

H

Hackerman, Norman, 192
Hallgren, Richard E., 44
Health. *See also* Accidents; AIDS; Disease
  decisions regarding, 181–183
  diagnostic technology to predict, 183–186
  family planning and, 173–174
  and malpractice issues, 92, 93
  relationship between diet and, 81–83
  and use of chewing tobacco, 84–85
Health care
  evaluation of, 101–103
  expenditures for, 181–183
  for homeless and mentally ill people, 190–191
  impact of biological advances on, 135
  inequitability of, 191
Heaton, George R., Jr., 162
High blood pressure, 81–82
High school, 5. *See also* School
Highway improvement, 27–29
Hip fractures, 102
Hispanic Americans, 210. *See also* Minorities

HIV (human immunodeficiency virus), 190–192. See also AIDS
Homeless people
  children as, 110–112
  effect of screening for disease on, 190–191
Hong Kong University of Science and Technology, 160
Housing
  need for affordable, 39, 41, 111
  technological advances in construction of, 39–41
Hubble Space Telescope, 150
Human behavior, economic forecasts and, 122
Human chromosomes, 144–146
Human reproduction
  in developing countries, 172–174
  and infertility issues, 88–91
  and obstetrical care system, 91–93
Humanities courses, 223
Hunger, 4–5, 167–169
Hurricanes, 44–46
Hussein, Saddam, 37, 138

I

Iacocca, Lee, 223
Illness. See Disease; Health
Immunization programs, 170–172
In vitro fertilization, 88, 91
Incas, crops of, 175, 177
Industrial production, of weapons, 37–38
Industry, Japanese vs. U.S., 158–159, 163, 164
Infants. See also Children
  anencephalic, 186–189
  care of, 116
  death of newborn, 174
  of HIV-infected mothers, 191
Infertility, 88–91

Informed consent, HIV testing and, 191
Infrastructure
  effect of sea level changes on, 50
  maintenance of, 46–48
Injury. See Accidents
Insects
  development of trees that resist, 129
  knowledge regarding, 128–129
  pesticides against, 69–72
Insurance. See also Medical insurance
  impact of genetic testing on, 186
  medical malpractice, 92, 93
International technology, 157–160. See also Technology
Iodine deficiency, 168
Iraq, invasion of Kuwait by, 35
Irrigation agriculture, 67, 72–74

J

Japan
  high school education in, 5
  monitoring of U.S. research facilities by, 157
  policies and industry in, 162–164
  U.S. college students studying in, 159
Jaynes, Gerald David, 107
Jefferson, Thomas, 137
Job evaluation techniques, 112
John Paul II, Pope, 213
Jones, Howard W., Jr., 88
Jortberg, Robert F., 46

K

Kahneman, Daniel, 123
Kates, Robert W., 167
Kean, Thomas H., x, 16

L

Kesterson National Wildlife Refuge, 72
Kindergarten stress, 201–203
King, Martin Luther, Jr., 219, 221
Koop, C. Everett, 96
Kurtz, Robert B., 35
Kuwait, Iraqi invasion of, 35
Kyoto University, 159

L

Laboratories, random audits of, 196
Latchkey children, 116
Lead poisoning, 95
Lederman, Leon M., 11
Legislation
  equal-opportunity, 113
  product-liability, 148
Lenie, Pieter, 60
Levi-Montalcini, Rita, 217
Levy, Eugene H., 130
Lewin, Lawrence S., 117
Liberal arts students, 12, 13
Light pollution, 149–151
Likins, Peter W., 160

M

Madison, Bernard L., 208
Magnetic fusion, 140
Magnuson, John J., 57
Malnutrition, childhood, 168
Malpractice, medical, 92, 93
Management practices, 163–164
Mars, U.S. and Soviet missions to, 130–133
Mastroianni, Luigi, Jr., 146
Mathematics education, 208–211
McDonald, John C., 33
McNair, Ronald, 219
Medical care. *See* Health care
Medical ethics
  and experimental treatment for AIDS, 99
  choosing among health services and, 181–183
  diagnostic technology and, 183–186
  organ donations and, 187–189
  regarding infertility treatment, 90
Medical insurance, 118, 191
Medical malpractice, 92, 93
Mentally ill people, 190–191
Merton, Robert, 217
Michael, Robert T., 112
Microbes, 129
Microchip industry, 24, 26
Microorganisms, genetically engineered, 62–64
Miller, Heather, 96
Mineral extraction, from ocean, 136
Minorities. *See also* Black Americans
  availability of medical services for, 191
  needs assessment for Soviet, 165, 166
  representation in engineering of, 161
  in science and mathematics-related areas, 161, 210, 215, 216, 220
Morrison, David L., 138
Morrison, Philip, 203
Morrison, Phylis, 203
Mortality rates, 172, 174
Motorcycle injuries, 95
Motulsky, Arno G., 81

N

National Communications System, 34
National Science Week, 3, 6
National Security Council, 35
Nations, James D., 74
Natural disasters, preparation for, 44–46

INDEX

Natural gas, 56
Nelkin, Dorothy, 183
Nicotine, 84
Nuclear power, 56
Nurse-midwives, 92

O

Obesity, 81–82
Obstetrical care system, 91–93
Ocean, resources within, 136–138
Offshore drilling, 136
Oregon Health Decisions, 182
Organ donations, 187–189
Organisms, genetically engineered, 62–64
Osteoporosis, 81–82

P

Palmer, John L., 115
Parents, role in exposure to science, 5
Parfit, Michael, 60
Parker, Frank L., 65
Patrusky, Ben, ix, 9
Pay equality, gender gap in, 112–114
Pearson, Willie, Jr., 219
Peer review, 195, 196
Penguins, 60, 61
Pepino, 177
Periodontal disease, 84
Persian Gulf crisis, 138
Pesek, John, 67
Pesticides
  and alternative systems of agriculture, 67–69
  contradictions over use of, 69–72
  in drainage water, 73
  residues in food, 67, 71
Photosynthesis, 127
Physicians, malpractice issues for, 92, 93
Physics, 134

Political decisions, 13
Pollution
  air. *See* Greenhouse effect
  light, 149–151
  water, 67, 72–74
Popenoe, Hugh, 175
Poverty
  among Black Americans, 107, 108
  child care issues and, 115–117
  and programs to combat hunger, 168, 169
Pregnancy
  dangers of, 172–173
  HIV testing during, 190–191
  teenage, 148
Press, Frank, ix, 151
Professional societies, 159
Public opinion research, Soviet, 166
Public school. *See* School
Publishing standards, 196

R

Rabi, I.I., 13
Race relations, 107–109
Radioactive waste disposal, 65–67
Raven, Peter H., 133
Rayman, Paula, 216
Reactor sites, 65–67
Recombinant DNA technology
  advances in use of, 134
  genetic engineering with, 63, 64
  and genetic structure and change in forest organisms, 129
Religious Fundamentalists, 211, 213
Replication, of experiments, 195–196
Research
  awareness and involvement in international, 157–160
  cooperative, 164

federal support of, 152–154
involving experiments on
    animals, 192–194
medical effectiveness, 103
military vs. civilian, 153
misconduct or sloppy, 195–197
Research methods
    Japanese, 163–164
    Soviet, 165, 166
Risk communication, 14–16
Rocoto, 177
Roebling, John, 223
Roebling, Washington, 223
Rubenstein, Arthur H., 195

S

Saccocio, Damian M., 24
Safety belts, school bus, 21–23
Salaries, 112–114, 214
Saturated fat, 82
School. *See also* Education
    attendance of homeless children in, 111
    contact with physical world in, 204–205
    exposure and motivation for science study in, 5, 161
    method of teaching science in, 10
    readiness for, 202
    teaching about evolution in, 211–213
    volunteers in, 205–208
School bus safety, 21–23
Science
    careers in, 5, 215–218
    federal budget decisions regarding, 152–154
    impact of advances on historical developments, 134
    integrity in, 195–197
    minorities in, 161, 210, 215
    misconceptions about, 3, 18
    phobia in relating to, 9–11
    politicians and, 18
    products vs. process of, 9–10

women in, 161, 210, 215–218
Science communication
    enhancement of science literacy by, 11
    image of scientists transmitted by, 10
    sensationalism in, 15
Science students
    and decision to study for doctorate, 214–216
    social sciences and humanities for, 12
Scientists
    assisting politicians in decision making, 16–18
    black, 219–221
    competition among, 217
    human qualities of, 10
    integrity among, 195–197
    media's distorted image of, 6–8
    misconceptions about, 3, 4
    promotion of science literacy by, 11–13
    role in educating public, 17
Sea level changes, 48–51, 138
Sea turtles, 57–59
Seabed resources, 136–138
Segregation, racial, 108, 109
Selenium, 72, 73
Sematech, 164
Semiconductor industry, 24, 26, 164
Sensationalism, in science communication, 15
Sewall, Gilbert T., 205
Sex discrimination, 218
Sexual behavior, and spread of AIDS, 96–98
Sexually transmitted diseases, 147–148. *See also* AIDS
Shrimping industry, 58, 59
Shrubs, 128–129
Simon, Herbert A., 123, 141
Smokeless tobacco, 83–85
Smoking, chewing tobacco and, 84
Snuff, 84

INDEX

Soviet Union
  mission to Mars planned by, 130–133
  statistics and research method improvement for, 165–167
Space exploration, 130–133
Sparks, Robert D., 86
Stanford University, 159
Statistical analyses, courtroom, 120–122
Stern, Paul C., 165
Stever, H. Guyford, 157
Stock market trends, 123–124
Straf, Miron L., 120
Submarines, 137
Substance abuse
  alcohol, 86–88
  chewing tobacco, 83–85
  drug treatment programs for, 117–119
  and spread of AIDS, 96–98, 101
Sulfur dioxide emissions, 139
Sununu, John, 223
Superconductivity, 152
Supreme Court, U.S., on evolution, 211–213
Sussman, John M., 30

T

Tamarillo, 177
Tancredi, Laurence, 183
Taylor, Michael R., 69
Technological development, U.S. vs. international, 157–160
Technological disasters, 14, 15
Technology
  federal budget decisions regarding, 152–154
  international, 157–160
  media's distorted image of, 6–8
Technology research associations, Japanese, 163
Teenagers. *See also* Children
  pregnancy in, 148, 173
  sexual behavior of, 98
Telephone system, 33–35

Telescopes, 150
Temperature rise, due to greenhouse effect, 49, 55–56
Terrorism, 33–35
Thaler, Richard H., 122
Thier, Samuel O., 101
Tobacco, chewing, 83–85
Tourism
  in Antarctica, 60–61
  to sea floor, 137
Transplantation programs, anencephalic infants and, 187–189
Transportation issues
  airport, 30–33
  rising sea level and, 50
  school bus safety, 21–23
  traffic problems, 27–29
Trees, 127, 129. *See also* Forests
Tropical disease, 171
Turtle excluder devices (TEDs), 59
Turtles, sea, 57–59
Tversky, Amos, 123

U

Ulluco, 177
UNICEF, 170
United Nations
  food supply and food relief programs by, 169
  immunization programs of, 170, 171
  national disaster reduction programs of, 45
Universities. *See also* College education
  degress awarded in forestry and related fields in, 129
  funding for black, 221
  Japanese students in U.S., 159
  publish or perish pressures in, 196
Ureaformaldehyde foam insulation, 120

V

Vaccines, 170-172
Values, and view of risk, 15
Van Schilfgaarde, Jan, 72
Viking mission of 1976, 132
Vitamin A deficiency, 168
Volunteerism, educational, 205-208

W

Wage gap, gender, 112-114
Waste disposal
  radioactive, 65-67
  use of seabed for, 137
Water quality
  impact of agriculture on, 67, 72-73
  regulations, 74
Weapon production, 37-38
Weiss, Robin, 99
Welfare hotels, 110-111
West Antarctic Ice Sheet, 61
White, Robert M., 55
Wildlife, effects of irrigation agriculture on, 72, 73
Williams, Robin M., Jr., 107
Witnesses, expert, 120-122
Women
  dangers of pregnancy for, 172-173
  in engineering, 161
  equality of pay for, 112-114
  families headed by, 110
  HIV testing in pregnant, 190-192
  impact of family planning on health of, 173-174
  in science and mathematics-related areas, 210, 215-218, 220
Wootan, Charley V., 21
World Health Organization, 170

Y

Young, Frank, 17